AQA GCSE (9–1)
Physics
Student Book

Sandra Mitchell
Charles Golabek
Series editor: Ed Walsh

William Collins' dream of knowledge for all began with the publication of his first book in 1819.

A self-educated mill worker, he not only enriched millions of lives, but also founded a flourishing publishing house. Today, staying true to this spirit, Collins books are packed with inspiration, innovation and practical expertise. They place you at the centre of a world of possibility and give you exactly what you need to explore it.

Collins. Freedom to teach

HarperCollins Publishers
1 London Bridge Street
London SE1 9GF

Browse the complete Collins catalogue at
www.collins.co.uk

First edition 2016

10 9 8 7 6 5

© HarperCollins*Publishers* 2016

ISBN 978-0-00-815877-4

Collins® is a registered trademark of HarperCollins Publishers Limited

www.collins.co.uk

A catalogue record for this book is available from the British Library

Commissioned by Lucy Rowland and Lizzie Catford

Edited by Hamish Baxter

Project managed by Elektra Media Ltd

Copy edited by Jim Newall

Development edited by Gwynneth Drabble and Gillian Lindsey

Proofread by Jan Schubert, Jo Kemp, Cassie Fox and Laurice Suess

Typeset by Jouve India and Ken Vail Graphic Design

With thanks to contributing author Crispin Myerscough

Cover design by We are Laura

Printed by Grafica Veneta, S.p.A., Italy

Cover images © Shutterstock/Jurik Peter, Emilio Segre Visual Archives/American Institute Of Physics/Science Photo Library

Approval message from AQA

This textbook has been approved by AQA for use with our qualification. This means that we have checked that it broadly covers the specification and we are satisfied with the overall quality. Full details of our approval process can be found on our website.

We approve textbooks because we know how important it is for teachers and students to have the right resources to support their teaching and learning. However, the publisher is ultimately responsible for the editorial control and quality of this book.

Please note that when teaching the GCSE Physics course, you must refer to AQA's specification as your definitive source of information. While this book has been written to match the specification, it cannot provide complete coverage of every aspect of the course.

A wide range of other useful resources can be found on the relevant subject pages of our website: aqa.org.uk

ACKNOWLEDGEMENTS

The publishers gratefully acknowledge the permissions granted to reproduce copyright material in this book. Every effort has been made to contact the holders of copyright material, but if any have been inadvertently overlooked, the Publisher will be pleased to make the necessary arrangements at the first opportunity.

Chapter 1

p12 Alex Staroseltsev/Shutterstock, Christos Siatos/Shutterstock, Kzenon/ Shutterstock, Neil Mitchell/Shutterstock; p13 You can more/Shutterstock, Ivan Smuk/Shutterstock, Filip Fuxa/Shutterstock; p14 bunnyphoto/Shutterstock, oknoart/Shutterstock, john michael evan potter/Shutterstock; p16 Digital Storm/Shutterstock; p17 WICHAI WONGJONGJAIHAN/Shutterstock; p18 Maria Masich/Shutterstock, bikeriderlondon/Shutterstock; p19 Chris Jenner/Shutterstock; p21 Izf/Shutterstock, Vlacheslav Lopatin/Shutterstock; p23 sasipixel/Shutterstock; p26 rdonar/Shutterstock, alterfather/Shutterstock; p27 Image Point Fr/Shutterstock; p28 Valeriy Lebedev/Shutterstock; p31 NinaM/Shutterstock; p32 gravitylight.org; p33 zhu difeng/Shutterstock; p35 Michael WickShutterstock; p36 Steve Collender/Shutterstock; p37 Parinyabinsuk/Shutterstock, Harri Aho/Shutterstock; p38 Fouad A. Saad / Shutterstock; p45 Ivan Smuk/Shutterstock

Chapter 2

p46 TED KINSMAN/SCIENCE PHOTO LIBRARY, asadykov/Shutterstock, Gearstd/Shutterstock, StudioFI/Shutterstock; p47 Timothy Hodgkinson/Shutterstock, a_v_d/Shutterstock; p48 Minerva Studio/Shutterstock; p49 TED KINSMAN/SCIENCE PHOTO LIBRARY; p59 Chones/Shutterstock; p64 ANDREW LAMBERT PHOTOGRAPHY/SCIENCE PHOTO LIBRARY; p66 a_v_d/Shutterstock; p68 Lukasz Pajor/Shutterstock; p69 michaeljung/Shuttersstock, Markus Gebauer/Shutterstock; p70 cristi180884/Shutterstock, _LeS_/Shutterstock; p72 Lilyana Vynogradova/Shutterstock; p73 imagedb.com/Shutterstock, You can more/Shutterstock; p74 Kotkot32/Shutterstock; p75 CPM PHOTO/Shutterstock, SvedOliver/Shutterstock, Monkey Business Images/Shutterstock

Chapter 3

p82 Freer/Shutterstock, Zmaj88/Shutterstock, Bplanet/Shutterstock; p83 Kekyalyaynen/Shutterstock, Petr Malyshev/Shutterstock; p85 SpaceKris/Shutterstock; p86 bilha Golan/Shutterstock; p87 Hurst Photo/Shutterstock; p88 Waraphorn Aphai/Shutterstock; p89 practicalaction.org; p93 Kekyalyaynen/Shutterstock; p95 Tyler Olson/Shutterstock; p96 Everett Historical/Shutterstock; p99 Cebas/Shutterstock

Chapter 4

p114 Raymond Llewellyn/Shutterstock; p121 PUBLIPHOTO DIFFUSION/SCIENCE PHOTO LIBRARY; p124 CristinaMuraca/Shutterstock; p126 Stefania Arca/Shutterstock; p127 Giovanni Cancemi/Shutterstock; p131 Torquetum/Shutterstock, PHILIPPE PLAILLY/SCIENCE PHOTO LIBRARY

Chapter 5

p140 sippakorn/Shutterstock, richsouthwales/Shutterstock, Marius Pirvu/Shutterstock; p141 popvartem.com/Shutterstock, thieury/Shutterstock, Danshutter/Shutterstock; p143 Rustam Shanov/Shutterstock, p146 Fedor Selivanov/Shutterstock; p147 Mariusz Szczygiel/Shutterstock; p150 Harry how/Getty; p159 Oleksandr Kalinichenko/Shutterstock; p160 VOLKER STEGER/SCIENCE PHOTO LIBRARY; p162 mandritoiu/Shutterstock; p168 siloto/Shutterstock; p171 gielmichal/Shutterstock; p180 Georgios Kollidas/Shutterstock, ChameleonsEye/Shutterstock, Vorm in Beeld/Shutterstock; p181 Graeme Dawes/Shutterstock

Chapter 6

p190 SirinS/Shutterstock, Tom Tom/Shutterstock, CristinaMuraca/Shutterstock, Peter Bernik /Shutterstock; p191 isarescheewin/Shutterstock, Artography/Shutterstock; p192 Matt Cardy/Getty; p194 SirinS/Shutterstock; p196 Joshua Rainey Photography/Shutterstock; p197 asharkyu/Shutterstock, Ryan Jorgensen – Jorgo/Shutterstock; p204 Sabphoto/Shutterstock, Imageman/Shutterstock; p208 KonstantinChristian/Shutterstock, Monkey Business Images; p209 kalewa/Shutterstock; p216 Denise Lett/Shutterstock, Suttha Burawonk/Shutterstock; p217 ZEPHYR/SCIENCE PHOTO LIBRARY; p218 Anna Kaminska/Shutterstock; p219 JIANG HONGYAN/Shutterstock, Ivan Smuk/Shutterstock; p232 Image Point Fr/Shutterstock

Chapter 7

p242 shooarts/Shutterstock; p243 zhengzaishuru/Shutterstock; p244 ChameleonsEye/Shutterstock; p245 revers/Shutterstock; p251 Paul Velgos/Shutterstock; p254 Chatchai-Rombix/Shutterstock; p257 Olga Popova/Shutterstock; p261 goldenjack/Shutterstock; p264 wk1003mike/Shutterstock

Chapter 8

p274 Anatolii Vasilev/Shutterstock; p275 NASA, NASA; p276 NASA/JPL/MSSS, NASA; p277 NASA, NASA; p278 NASA, NASA; p279 NASA; p280 NASA, NASA, NASA/JPL-Caltech; p281 Vadim Nefedoff/Shutterstock, science@NASA; p282 NASA, NASA/SDO/AIA; p284 NASA, ESA, C.R. O'Dell (Vanderbilt University), and M. Meixner, P. McCullough, p285 NASA; p290 NASA/SCIENCE PHOTO LIBRARY, Arthur Balitskiy/Shutterstock; p292 NASA/ESA; p293 NASA

Contents

You can use this book if you are studying Combined Science: Trilogy

 you will need to master all of the ideas and concepts on these pages

 you will need to master some of the ideas and concepts on these pages.

How to use this book   6

Chapter 1   Energy   12
- 1.1   Potential energy   14
- 1.2   Investigating kinetic energy   16
- 1.3   Work done and energy transfer   18
- 1.4   Understanding power   20
- 1.5   Specific heat capacity   22
- 1.6   Required practical: Investigating specific heat capacity   24
- 1.7   Dissipation of energy   26
- 1.8   Energy efficiency   28
- 1.9   Required practical: Investigating ways of reducing the unwanted energy transfers in a system   30
- 1.10   Using energy resources   32
- 1.11   Global energy supplies   34
- 1.12   Key concept: Energy transfer   36
- 1.13   Maths skills: Calculations using significant figures   38
- 1.14   Maths skills: Handling data   40

Chapter 2   Electricity   46
- 2.1   Static electricity   48
- 2.2   Electric fields   50
- 2.3   Electric current   52
- 2.4   Series and parallel circuits   54
- 2.5   Investigating circuits   56
- 2.6   Circuit components   58
- 2.7   Required practical: Investigate, using circuit diagrams to construct circuits, the I–V characteristics of a filament lamp, a diode and a resistor at constant temperature   60
- 2.8   Required practical: Use circuit diagrams to set up and check appropriate circuits to investigate the factors affecting the resistance of electrical circuits, including the length of a wire at constant temperature and combinations of resistors in series and parallel   62
- 2.9   Control circuits   64
- 2.10   Electricity in the home   66
- 2.11   Transmitting electricity   68
- 2.12   Power and energy transfers   70
- 2.13   Calculating power   72
- 2.14   Key concept: What's the difference between potential difference and current?   74
- 2.15   Maths skills: Using formulae and understanding graphs   76

Chapter 3   Particle model of matter   82
- 3.1   Density   84
- 3.2   Required practical: To investigate the densities of regular and irregular solid objects and liquids   86
- 3.3   Changes of state   88
- 3.4   Internal energy   90
- 3.5   Specific heat capacity   92
- 3.6   Latent heat   94
- 3.7   Particle motion in gases   96
- 3.8   Increasing the pressure of a gas   98
- 3.9   Key concept: Particle model and changes of state   100
- 3.10   Maths skills: Drawing and interpreting graphs   102

Chapter 4   Atomic structure   108
- 4.1   Atomic structure   110
- 4.2   Radioactive decay   112
- 4.3   Background radiation   114
- 4.4   Nuclear equations   116
- 4.5   Radioactive half-life   118
- 4.6   Hazards and uses of radiation   120
- 4.7   Irradiation   122
- 4.8   Uses of radiation in medicine   124
- 4.9   Using nuclear radiation   126
- 4.10   Nuclear fission   128
- 4.11   Nuclear fusion   130
- 4.12   Key concept: Developing ideas for the structure of the atom   132
- 4.13   Maths skills: Using ratios and proportional reasoning   134

Chapter 5   Forces   140
- 5.1   Forces   142
- 5.2   Speed   144
- 5.3   Acceleration   146
- 5.4   Velocity–time graphs   148
- 5.5   Calculations of motion   150

5.6	Heavy or massive?	152
5.7	Forces and motion	154
5.8	Resultant forces	156
5.9	Forces and acceleration	158
5.10	Required practical: Investigating the acceleration of an object	160
5.11	Newton's third law	162
5.12	Momentum	164
5.13	Keeping safe on the road	166
5.14	Moments	168
5.15	Levers and gears	170
5.16	Pressure in a fluid	172
5.17	Atmospheric pressure	174
5.18	Forces and energy in springs	176
5.19	Required practical: Investigate the relationship between force and the extension of a spring	178
5.20	Key concept: Forces and acceleration	180
5.21	Maths skills: Making estimates of calculations	182

Chapter 6 Waves | **190** |
6.1	Describing waves	192
6.2	Transverse and longitudinal waves	194
6.3	Key concept: Transferring energy or information by waves	196
6.4	Measuring wave speeds	198
6.5	Required practical: Measuring the wavelength, frequency and speed of waves in a ripple tank and waves in a solid	200
6.6	Reflection and refraction of waves	202
6.7	Required practical: Investigate the reflection of light by different types of surface and the refraction of light by different substances	204
6.8	Sound waves	206
6.9	Exploring ultrasound	208
6.10	Seismic waves	210
6.11	The electromagnetic spectrum	212
6.12	Reflection, refraction and wavefronts	214
6.13	Gamma rays and X-rays	216
6.14	Ultraviolet and infrared radiation	218
6.15	Required practical: Investigate how the amount of infrared radiation absorbed or radiated by a surface depends on the nature of that surface	220

6.16	Microwaves	222
6.17	Radio and microwave communication	224
6.18	Colour	226
6.19	Lenses	228
6.20	Images and magnification	230
6.21	Emission and absorption of infrared radiation	232
6.22	Temperature of the Earth	234
6.23	Maths skills: Using and rearranging equations	236

Chapter 7 Electromagnetism | **242** |
7.1	Magnetism and magnetic forces	244
7.2	Compasses and magnetic fields	246
7.3	The magnetic effect of a solenoid	248
7.4	Electromagnets in action	250
7.5	Calculating the force on a conductor	252
7.6	Electric motors	254
7.7	Loudspeakers	256
7.8	The generator effect	258
7.9	Key concept: The link between electricity and magnetism	260
7.10	Using the generator effect	262
7.11	Transformers	264
7.12	Maths skills: Rearranging equations	266

Chapter 8 Space | **274** |
8.1	The Solar System	276
8.2	Orbits of planets, moons and artificial satellites	278
8.3	The Sun and other stars	280
8.4	Main sequence of a star	282
8.5	Life cycles of stars	284
8.6	How the elements are formed	286
8.7	Red-shift	288
8.8	Key concept: Gravity: the force that binds the Universe	290
8.9	Maths skills: Using scale and standard form	292

Appendix–Circuits | **298** |
Appendix–Equations | **299** |
Glossary | **301** |
Index | **311** |

How to use this book

Remember! To cover all the content of the AQA Physics Specification you should study the text and attempt the End of chapter questions.

These tell you what you will be learning about in the lesson and are linked to the AQA specification.

This introduces the topic and puts the science into an interesting context.

Each topic is divided into three sections. The level of challenge gets harder with each section.

Each section has level-appropriate questions, so you can check and apply your knowledge.

Physics

Radioactive half-life

KEY WORD

half-life

Learning objectives:
- explain what is meant by radioactive half-life
- calculate half-life
- calculate the decline in activity after a number of half-lives.

Radioisotopes are often used to monitor a biological process in the body or in the environment. We need to know the half-lives of different radioisotopes so that we can minimise contamination or irradiation.

Half-life

We cannot predict when the nucleus of one particular atom will decay. It could be next week or not for a million years. Radioactive decay is a random process. If there is a very large number of atoms, some of them will decay each second. We plot a graph of activity against time and draw a curve of best fit. We can then use this curve to find when the activity has halved (Figure 4.12). The **half-life** of a radioisotope is the average time it takes for half the nuclei present to decay, or the time it takes for the activity to fall to half its initial level. We use half-life because we cannot predict the time it will take for all the atoms to decay.

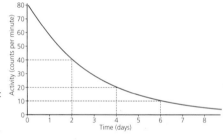

Figure 4.12 The time it takes for the activity to halve is constant

The activity of a radioactive substance gets less and less as time goes on. The graph line in Figure 4.12 gets closer and closer to the time axis but never reaches it because the activity halves each half-life.

1. Explain why you can't predict when a particular atom will decay.
2. Explain what is meant by 'half-life'.
3. What will the activity in Figure 4.12 be after:
   a 2 days? **b** 4 days? **c** 6 days? **d** 8 days?

Calculating half-life

To calculate the half-life of a radioisotope, plot a graph of count rate detected (which is proportional to the total activity) against time, as shown in Figure 4.13. The background count should be subtracted from each reading before the graph is plotted. Plot the points and then draw a smooth curve of best fit through the points. Then find several values for the half-life from the graph by finding the time for the activity in counts per minute to fall from 80 to 40, 40 to 20, 20 to 10 and so on. You should calculate the average of the values you have found. The time for the activity to halve may not be exactly the same each time. Any differences will be due to the random nature of radioactive decay.

KEY INFORMATION

When you draw a curve of best fit on a graph of activity against time for a radioisotope, not all of the points will be on the curve. Some will be above the curve and some will be below the curve because of the random nature of radioactive decay.

118　AQA GCSE Physics: Student Book

---

④ Calculate the half-life of the radioisotope shown in Figure 4.13.

4.5

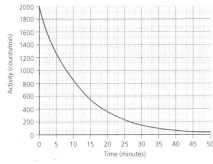

Figure 4.13

⑤ The table shows the activity of a radioactive sample over time.

a Draw a graph of activity against time.

b Draw a best-fit curve through the points.

c Calculate the half-life by looking at time for the activity to halve at three different points on your curve. Calculate the average.

⑥ The activity of a radioactive sample took 4 hours to decrease from 100 Bq to 25 Bq. Calculate its half-life.

Time in minutes	Activity in Bq
0	100
0.5	76
1.0	51
1.5	40
2.0	26
2.5	18
3.0	12
3.5	10
4.0	8

HIGHER TIER ONLY

Calculating decline in activity

For some applications such as a smoke alarm, a radioisotope with a long half-life is most suitable so that the rate of decay does not decrease significantly. Some radioisotopes are used as tracers, using the radiation they emit to trace the path of a substance the radiosope is attached to.

Radioisotopes can be used as environmental tracers (such as detecting a leak in a pipe) or medical tracers. For these applications, a short half-life is best. This means that the activity will decrease to a level similar to the background count fairly quickly because the time taken for half the radioactive nuclei to decay is very short.

⑦ Technetium-99m is widely used in medicine. It has a half-life of 6 hours. How much of it remains after 1 day?

⑧ What fraction of the original sample remains after 80 minutes in Figure 4.13?

⑨ Caesium-134 has a half-life of 2 years. If its activity was monitored for 6 years, what fraction of the original activity level would it have dropped to after 6 years?

DID YOU KNOW?

The half-lives of different radioactive isotopes vary from a fraction of a second to millions of years.

Google search: 'half-life'   119

The first page of a chapter has links to ideas you have met before, which you can now build on.

This page gives a summary of the exciting new ideas you will be learning about in the chapter.

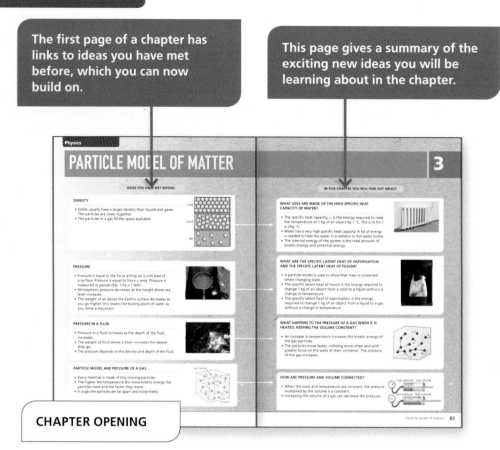

CHAPTER OPENING

The Key Concept pages focus on a core ideas. Once you have understood the key concept in a chapter, it should develop your understanding of the whole topic.

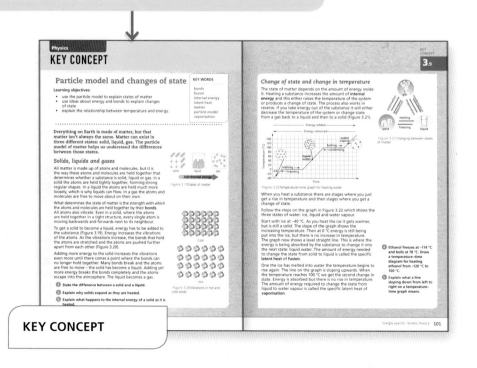

KEY CONCEPT

There is a dedicated page for every Required Practical in the AQA specification. They help you to analyse the practical and to answer questions about it.

The tasks – which get a bit more difficult as you go through – challenge you to apply your science skills and knowledge to the new context.

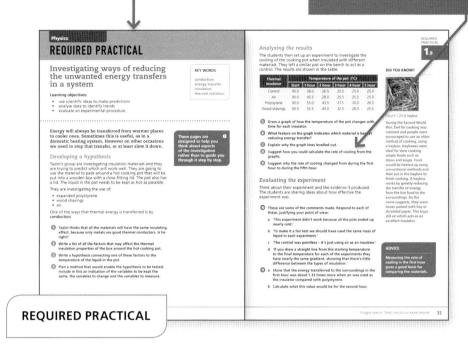

REQUIRED PRACTICAL

The Maths Skills pages focus on the maths requirements in the AQA specification, explaining concepts and providing opportunities to practise.

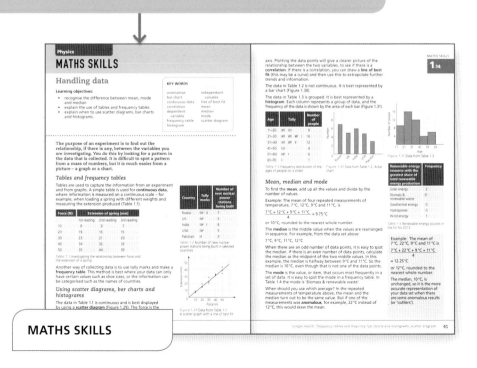

MATHS SKILLS

Physics

These lists at the end of a chapter act as a checklist of the key ideas of the chapter. In each row, the green box gives the ideas or skills that you should master first. Then you can aim to master the ideas and skills in the blue box. Once you have achieved those you can move on to those in the red box.

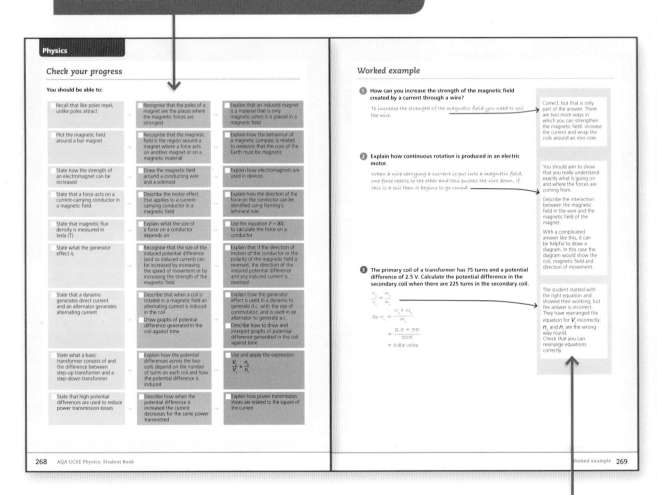

Use the comments to help you understand how to answer questions. Read each question and answer. Try to decide if, and how, the answer can be improved. Finally, read the comments and try to answer the questions yourself.

END OF CHAPTER

The End of chapter questions allow you and your teacher to check that you have understood the ideas in the chapter, can apply these to new situations, and can explain new science using the skills and knowledge you have gained. The questions start off easier and get harder. If you are taking Foundation tier, try to answer all the questions in the Getting started and Going further sections. If you are taking Higher tier, try to answer all the questions in the Going further, More challenging and Most demanding sections.

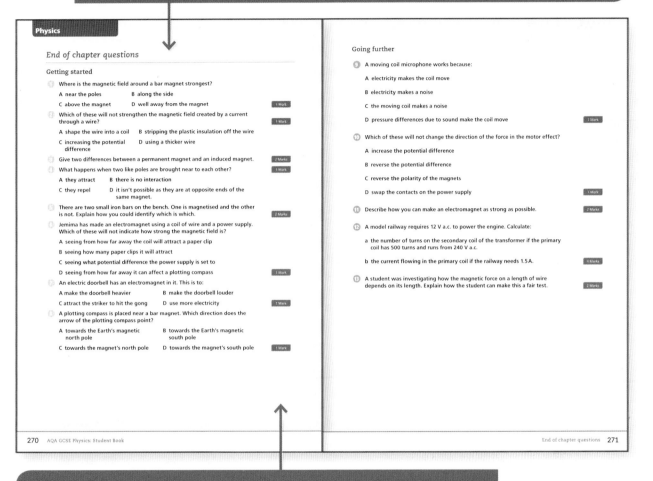

End of chapter questions

Getting started

1. Where is the magnetic field around a bar magnet strongest?

   A near the poles   B along the side

   C above the magnet   D well away from the magnet   1 Mark

2. Which of these will not strengthen the magnetic field created by a current through a wire?   1 Mark

   A shape the wire into a coil   B stripping the plastic insulation off the wire

   C increasing the potential difference   D using a thicker wire

3. Give two differences between a permanent magnet and an induced magnet.   2 Marks

4. What happens when two like poles are brought near to each other?   1 Mark

   A they attract   B there is no interaction

   C they repel   D it isn't possible as they are at opposite ends of the same magnet.

5. There are two small iron bars on the bench. One is magnetised and the other is not. Explain how you could identify which is which.   2 Marks

6. Jemima has made an electromagnet using a coil of wire and a power supply. Which of these will not indicate how strong the magnetic field is?

   A seeing from how far away the coil will attract a paper clip

   B seeing how many paper clips it will attract

   C seeing what potential difference the power supply is set to

   D seeing from how far away it can affect a plotting compass   1 Mark

7. An electric doorbell has an electromagnet in it. This is to:

   A make the doorbell heavier   B make the doorbell louder

   C attract the striker to hit the gong   D use more electricity   1 Mark

8. A plotting compass is placed near a bar magnet. Which direction does the arrow of the plotting compass point?

   A towards the Earth's magnetic north pole   B towards the Earth's magnetic south pole

   C towards the magnet's north pole   D towards the magnet's south pole   1 Mark

Going further

9. A moving coil microphone works because:

   A electricity makes the coil move

   B electricity makes a noise

   C the moving coil makes a noise

   D pressure differences due to sound make the coil move   1 Mark

10. Which of these will not change the direction of the force in the motor effect?

    A increase the potential difference

    B reverse the potential difference

    C reverse the polarity of the magnets

    D swap the contacts on the power supply   1 Mark

11. Describe how you can make an electromagnet as strong as possible.   2 Marks

12. A model railway requires 12 V a.c. to power the engine. Calculate:

    a the number of turns on the secondary coil of the transformer if the primary coil has 500 turns and runs from 240 V a.c.

    b the current flowing in the primary coil if the railway needs 1.5 A.   4 Marks

13. A student was investigating how the magnetic force on a length of wire depends on its length. Explain how the student can make this a fair test.   2 Marks

There are questions for each assessment object (AO) from the final exams. This will help you develop the thinking skills you need to answer each type of question:

AO1 – to answer these questions you should aim to **demonstrate** your knowledge and understanding of scientific ideas, techniques and procedures.

AO2 – to answer these questions you should aim to **apply** your knowledge and understanding of scientific ideas and scientific enquiry, techniques and procedures.

AO3 – to answer these questions you should aim to **analyse** information and ideas to: interpret and evaluate, make judgements and draw conclusions, and develop and improve experimental procedures.

ENERGY

WE CAN MEASURE HOW MUCH ENERGY IS TRANSFERRED AND HOW QUICKLY

- Energy can be stored and also transferred from one store to another. Energy changes are measured in joules (J) or kilojoules (kJ).
- The energy values of different foods can be compared.
- The power ratings of appliances are measured in watts (W) or kilowatts (kW).

TEMPERATURE AND ENERGY

- Temperature tells us how hot something is.
- When there is a difference in temperature between two objects, energy is transferred from the hotter object to the colder one.
- Energy transfer tends to reduce the temperature difference.

TRANSFER OF THERMAL ENERGY

- Thermal energy is transferred by conduction, convection and radiation.
- In conduction and convection, energy is transferred by the movement of particles.
- Radiation is the only way energy can be transferred in a vacuum.

FOSSIL FUELS AND ALTERNATIVE ENERGY RESOURCES

- Fossil fuels are burnt to release the energy stored in them.
- Fossil fuels were formed over millions of years and supplies are running out.
- There are alternative energy resources which have advantages and disadvantages.

IN THIS CHAPTER YOU WILL FIND OUT ABOUT:

WHAT IS THE CONNECTION BETWEEN ENERGY TRANSFER AND POWER?

- Energy is transferred by heating, by electric current in a circuit, and when work is done by a force.
- We can measure the rate at which energy is being transferred or the rate at which work is done – this is called power.

WHAT IS THE CONNECTION BETWEEN ENERGY CHANGES AND TEMPERATURE CHANGE?

- We can calculate the energy stored in or released from a system when its temperature changes.
- The rate of cooling of a building is affected by the thickness and the thermal conductivity of its walls. Insulation can be used to reduce the transfer of energy by conduction and convection.

HOW CAN WE MONITOR AND CONTROL THE TRANSFER OF ENERGY?

60 J

0.2 J useful energy to give light

59.8 J energy heats the air

- Energy can be transferred usefully, stored or dissipated. The total amount of energy does not change. We can calculate the energy efficiency for any energy transfer. Some energy transfers are wasteful; we can try to reduce them.

WHAT IS THE ENVIRONMENTAL IMPACT OF DIFFERENT ENERGY RESOURCES?

- Fuels such as coal, oil, gas and nuclear fuel are not renewable. Supplies will run out.
- The use of fossil fuels is changing as more renewable energy resources are used for transport, electricity generation and heating.
- Most renewable resources do not generate a predictable (reliable) amount of electrical power.
- There are different environmental issues for each different energy resource.

Potential energy

Learning objectives:

- consider what happens when a spring is stretched
- describe what is meant by gravitational potential energy
- calculate the energy stored by an object raised above ground level.

KEY WORDS

elastic potential
 energy
gravitational
 field strength
gravitational
 potential energy

On Jupiter, gravity is three times stronger than on Earth. If the same mass was lifted to the same height on Earth and Jupiter, on Jupiter it would store three times the gravitational potential energy.

Stored energy

When a spring is stretched it stores energy; for example, in the stretching of a catapult. We call this **potential energy**. The stretched spring stores **elastic potential energy**. It stretches more if a greater force is applied, and returns to its original length when the force is removed.

Compressed springs are also used to store elastic potential energy; for example, in a wind-up clockwork toy.

Figure 1.1 A stretched catapult stores elastic potential energy

1 **Imagine slowly stretching a rubber band. Describe what you would feel as you stretch it more.**

2 **Suggest what happens to the amount of energy stored in the spring inside a the toy in Figure 1.2 as the key is turned.**

Water stored behind a dam also stores potential energy. It is called **gravitational potential energy (GPE)**. The GPE stored by an object can be increased by moving it upwards. For example, you gain GPE by going up stairs.

Figure 1.2 Turning the key twists a spring inside the toy. This stores elastic potential energy which can be transferred to kinetic energy

3 **An aircraft is flying horizontally at a height of 10 000 m. Explain whether its gravitational potential energy is increasing, decreasing or remaining constant.**

Calculating changes in gravitational potential energy

Tom lifts a box. The amount of gravitational potential energy gained by the box depends on:

- the mass of the box, m, in kilograms, kg
- the height Tom raises the box, h, in metres, m.

We can calculate the amount of gravitational potential energy gained (E_p) using the equation:

$E_p = mgh$

Figure 1.3 The stored gravitational potential energy of water held behind a dam can be transferred to kinetic energy if the water is released

where m is measured in kg, g is the **gravitational field strength** in N/kg and h is measured in m.

The pull of gravity on the box (its weight) is calculated from the equation:

weight = mass × gravitational field strength, $W = m \times g$

The gravitational field strength is the pull of gravity on each kilogram. The value of g is 9.8 N/kg at the Earth's surface. Changes in E_p are measured in joules (J).

Zack gains gravitational potential energy when he walks up some stairs (Figure 1.4). The E_p Zack gains can be calculated using $E_p = mgh$ where h is the vertical height he raises his body.

Zack then walks on a level floor. He does not gain or lose any gravitational potential energy now because his height above the floor does not change.

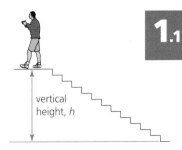

Figure 1.4 Vertical height, h

4 Lars is a weightlifter. He lifts a mass of 300 kg through a height of 2 m. Calculate the gravitational potential energy gained by the mass.

5 Sian picks up a ball from the floor and holds it 2 m above the ground. The ball has a mass of 60 g. Calculate the gravitational potential energy gained by the ball.

Calculating elastic potential energy

The amount of elastic potential energy stored in a spring can be increased by:

- increasing the extension of the spring, e, in metres, m
- increasing the spring constant, k, in newtons per metre, N/m.

We can calculate the elastic potential energy, E_e, stored in a spring using the equation:

$$E_e = \frac{1}{2}ke^2$$

Example: Calculate the energy stored in a spring when it is extended by 6 cm. The spring constant is 150 N/m.

Answer: 6 cm = 0.06 m

$E_e = 0.5 \times 150 \text{ N/m} \times (0.06 \text{ m})^2$

$\quad = 0.27 \text{ J}$

6 Calculate the energy stored in a spring which has a spring constant of 300 N/m and is extended by 0.1 m.

7 A spring has a spring constant of 500 N/m. The original length of the spring is 20 cm. It is stretched to a length of 25 cm. Calculate the energy stored in the spring.

8 A spring stores 12 J when it is stretched by 16 cm. Calculate the spring constant.

9 A stretched spring has a total length of 20 cm and a spring constant of 200 N/m. It is storing 0.25 J in its elastic potential energy store. Determine the unstretched length of the spring.

DID YOU KNOW?

A new pumped storage hydropower project in Chile will use solar power to pump water from the Pacific Ocean 600 metres above sea level to the top of a cliff. The total reservoir capacity of about 55 million cubic metres gives a maximum total energy stored of about 3.3 × 10^14 J. However, the amount of water that can be pumped by solar power is only 4.5 million cubic metres per day.

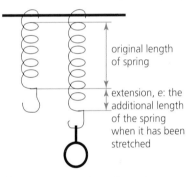

original length of spring

extension, e: the additional length of the spring when it has been stretched

Figure 1.5 Extension of a spring

MATHS

Always put values into an equation using SI units (time in seconds, distance in metres etc.). If the values in the question are not in SI units, you have to change the values to SI units before starting the calculation.

REMEMBER!

Use the **vertical** height when finding the change in an object's height.

Investigating kinetic energy

Learning objectives:

- describe how the kinetic energy store of an object changes as its speed changes
- calculate kinetic energy
- consider how energy is transferred.

Formula One racing cars reach speeds of 300 km/h or more. This means the kinetic energy store is very large. If the car is involved in a collision, this energy store may be reduced to zero very rapidly, causing lots of damage.

Kinetic energy

Energy must be transferred to make things move. A car uses energy from petrol or diesel to move. The greater its mass and the faster it goes, the more energy is transferred to the car's **kinetic energy** store and the higher the rate at which fuel is used.

Figure 1.6 The engine applies a force that transfers energy from a store in the fuel to a store of kinetic energy

1. **Why does an adult have more kinetic energy than a child when running at the same speed?**

2. **What is the 'fuel' for a child running around a playground?**

Calculating kinetic energy

The kinetic energy store of a moving object can be increased by:

- increasing the mass of the object, m in kg
- increasing its speed, v in m/s.

We can calculate the kinetic energy (E_k) of an object by using this equation:

$$E_k = \frac{1}{2} mv^2$$

Example: A car of mass 1600 kg is travelling at a steady speed of 10 m/s.

a Calculate the car's kinetic energy.

b The car's speed increases to 20 m/s. Calculate how much its kinetic energy store increases.

  a E_k at 10 m/s $= \frac{1}{2} mv^2 = \frac{1}{2} \times 1600 \text{ kg} \times (10 \text{ m/s})^2$
  $= \frac{1}{2} \times 1600 \times 100 = 80\,000$ J

  b E_k at 20 m/s $= \frac{1}{2} mv^2 = \frac{1}{2} \times 1600 \text{ kg} \times (20 \text{ m/s})^2$
  $= \frac{1}{2} \times 1600 \times 400 = 320\,000$ J

The increase in the car's $E_k = 320\,000 - 80\,000 = 240\,000$ J.

Doubling the speed has increased the car's E_k by a factor of 4. Note that speed is squared in the equation for E_k.

DID YOU KNOW?

Hollywood star Idris Elba smashed an 88-year-old record by driving at an average speed of 180.361 mph over a measured mile across the Pendine Sands in Wales in May 2015. The award-winning star beat the 1927 record set by Sir Malcolm Campbell.

3 a Meena is riding her bicycle at 2 m/s. She has a mass of 50 kg. Calculate her kinetic energy.

   b Meena doubles her speed. What happens to her kinetic energy?

We can convert from km/h to m/s by:

- multiplying by 1000 (there are 1000 m in a km)
- dividing by 3600 (there are 3600 seconds in an hour).

For example, $300 \text{ km/h} = \dfrac{(300 \times 1000 \text{ m})}{3600 \text{ s}} = 83.3 \text{ m/s}$

4 A car of mass 1200 kg increases its speed from 10 m/s to 30 m/s. By how much has its E_k increased?

5 Change a speed of 240 km/h to m/s.

Dropping a ball

If a ball is held 2 m above the ground, it has **gravitational potential energy** relative to a ball on the ground. Allowing it to fall transfers energy from a gravitational potential energy store to a kinetic energy store as it drops. The ball hits the ground and bounces. As it moves upwards, the kinetic **energy store** decreases as energy is transferred back to the gravitational potential energy store. The ball does not return to its original height though, because some of the energy is transferred to the surroundings by heating.

Figure 1.7 Time-lapse photo showing the path of a ball that is dropped

6 At what point does the ball in the photo:

   a have most gravitational potential energy?

   b have most kinetic energy?

7 Describe the changes to energy stores when a ball is thrown upwards.

8 A ball with a mass of 2 kg is dropped from a height of 20 m. Assuming all of the GPE transfers to E_k, calculate the speed of the ball as it hits the ground (g = 9.8 N/kg).

Work done and energy transfer

Learning objectives:

- understand what is meant by work done
- explain the relationship between work done and force applied
- identify the transfers between energy stores when work is done against friction.

People on a roller coaster experience rapid energy changes and experience G-forces similar to those experienced by astronauts.

Work done by a force

In science, **work** is only done by a **force** when an object moves.

More work is done when:

- the force is bigger
- the object moves further – its displacement is bigger.

Sam's car has broken down. He tries to push it but the car does not move. He is not doing any work though. Work is only done when a force *moves*. Sam gets some friends to help. Together, they can push with a larger force and the car moves. They are all doing work.

1 **What affects the amount of work done by a force?**

2 **What force moves when someone jumps off a wall?**

Calculating work done

The equation that links work, force and distance is:

Work done = force × distance moved along the line of action of the force

$W = F\,s$

where work done, W, is in joules, force, F, is in newtons and distance, s, is in metres.

When a person climbs stairs or jumps in the air the force moved is their weight.

Example: Dev climbs a flight of stairs rising a vertical height of 5 m. He weighs 600 N. Calculate the work Dev does by lifting his weight up the stairs.

Work done = force × distance moved along the line of action of the force

Work done = 600 N × 5 m = 3000 J

3 **A gymnast (Figure 1.9) weighs 400 N. How much work does she do when she jumps from the ground onto a beam 1.5 m above the ground?**

Figure 1.8 These men are doing work because the car is moving

Figure 1.9 The gymnast does work when she jumps up onto the beam from the floor

4. Mia is holding a 20 N weight without moving. How much work is she doing?

5. Amrita does 300 J of work in lifting a box with a force of 200 N. How high does she lift it?

Energy calculations

When work is done on an object there can also be a change in its **kinetic energy**. We can use this to calculate the force needed to stop a car when the distance it travels while coming to rest is known.

Example: A car of mass 1000 kg does an emergency stop when travelling at 15 m/s. It stops in a distance of 20 m. Calculate the braking force.

$$E_k \text{ of car} = \frac{1}{2}mv^2$$

$$= \frac{1}{2} \times 1000 \text{ kg} \times (15 \text{ m/s})^2$$

$$= 112\,500 \text{ J}$$

Work has to be done to reduce the kinetic energy of the car and bring it to a stop. The force that does the work is the friction force between the brakes and the wheel. The work done by the braking force is 112 500 J.

$W = F\,s$, where F is the braking force and s is the distance moved during braking. We can rearrange this to make F the subject of the equation.

$$F = \frac{W}{s}$$

$$= \frac{112\,500 \text{ J}}{20 \text{ m}}$$

$$= 5625 \text{ N}$$

Work done against the frictional forces acting on an object causes a rise in the temperature of the object. The temperature of the brakes increases.

6. Tom does 3000 J of work against friction in pushing a small van a distance of 12 m. How big is the friction force he has to push against?

7. A car of mass 800 kg is travelling at 12 m/s.

   a Calculate its kinetic energy.

   b The brakes are applied. What force from the brakes is needed if the car stops after travelling 8 m?

8. A pole vaulter has a weight of 500 N.

   a She vaults to a height of 4 m. How much work does she do?

   b How much kinetic energy does she have just before she lands?

   c When she lands, her trainers compress by 1 cm. Calculate the average force acting on her trainers as she is landing.

DID YOU KNOW?

A water-powered inclined railway has no engine. The top car has a full water tank which makes it slightly heavier than the car at the bottom. The two cars are attached by a cable going over a pulley. The extra weight of the top car does work to lift the lower car up the slope. Work is also done to overcome the force of friction.

Figure 1.10 Inclined railway

MAKING LINKS

You will need to link the information given here to the forces topic 5.13, where you will look at the factors that affect stopping distance.

Understanding power

Learning objectives:

- define power
- compare the rate of energy transfer by various machines and electrical appliances
- calculate power.

A human being can be considered an energy transfer device. Our energy comes from our food, about 10 MJ (10 million joules) every day.

Power

Imagine a tall office block with two lifts of the same mass. One lift takes 40 s to go up to the tenth floor; the other, newer lift, takes 25 s. Both lifts do the same amount of work but the newer one does it more quickly. The transfer of energy is more rapid; the new lift has more **power**.

Power is the rate of doing work or transferring energy. A machine that is more powerful than another machine transfers more energy each second.

Power is measured in watts (W). If one joule of energy is transferred in one second, this is one watt of power. 1 W = 1 J/s.

The table shows the typical power of various electrical appliances.

	Power / W
Kettle	2500
Microwave oven	1100
Iron	1000
Hairdryer	2000
Vacuum cleaner	1600
Television	114
Food blender	150

1 Which appliance is the most powerful?

2 Explain why a television might do more work than a food blender.

Calculating power

$$\text{Power} = \frac{\text{work done in J}}{\text{time in s}} \text{ or } \frac{\text{energy transferred in J}}{\text{time in s}}$$

These formulae can be written as:

$$P = \frac{W}{t} \text{ and}$$

$$P = \frac{E}{t}$$

Example: A machine does 1000 J of work in 8 seconds. What power does it develop?

$$\text{Power} = \frac{\text{work done}}{\text{time}} = \frac{1000\,\text{J}}{8\,\text{s}} = 125\,\text{W}$$

ADVICE

Energy is measured in joules (J) and power in watts (W). An energy transfer of 1 joule per second is equal to a power of 1 watt.

3 A toaster transferred **108 000 J** of energy in **2 minutes**. What is the power of the toaster?

4 An electric kettle is rated at **2 kW (2000 W)**. How much energy is transferred in **30 s**?

5 A crane lifts a load weighing **4000 N** through a height of **6 m** in **20 s**.

   a How much work does the crane do to lift the load, assuming there is no friction?

   b What is the power of the crane?

Personal power

Mel decides to work out her leg power. She runs up 16 stairs, each 20 cm high, in 10 s (Figure 1.11). Mel's mass is 50 kg.

Work done = force × distance moved in the direction of the force.

The force moved is Mel's weight. Assume g = 9.8 N/kg.

Weight = 50 × 9.8 = 490 N.

The **vertical** height of the stairs (in m) = 16 × 0.2 = 3.2 m.

Work done = 490 N × 3.2 m = 1568 J.

$$\text{Power} = \frac{\text{work done}}{\text{time taken}} = \frac{1568\,\text{J}}{10\,\text{s}} = 156.8\,\text{W}$$

Figure 1.11 How can Mel measure the vertical height?

6 Al weighs **800 N**. He climbs **5 m** vertically in **10 s** when he runs up the stairs of an office block.

   a How much work does he do?

   b Calculate his power.

7 The Eiffel Tower in Paris (Figure 1.12) is **300 m** high. Louis took **15 minutes** to climb its **1792 steps**. He has a mass of **60 kg**. What was his average power output during the climb?

8 Emma decides to measure her personal leg power doing step-ups. The step is **10 cm** high. Emma does **20 step-ups** in **30 s**. Her mass is **60 kg**. Calculate her leg power.

Figure 1.12 The Eiffel Tower

9 A car is moving at **108 km/h** and the engine provides a constant force of **1000N**. Calculate:

   a the distance the car moves in 1 second

   b the power of the engine.

Specific heat capacity

Learning objectives:

- understand how things heat up
- find out about heating water
- find out about specific heat capacity.

If you eat a jam sponge pudding soon after taking it out of the oven the jam seems to be hotter than the sponge, even though they have been cooked at the same temperature. This is because at the same temperature the jam stores more thermal energy than the pudding around it.

Hot and cold

It takes more energy to get some things hot than others. Different materials need different amounts of energy to raise the temperature by a given amount. Imagine a 1 kg block of copper and a 1 kg block of steel. From experiments we can show that:

- it takes 380 J of energy to raise the temperature of 1 kg of copper by 1 °C
- it takes 450 J of energy to raise the temperature of 1 kg of steel by 1 °C.

The amount of energy needed to change the temperature of an object depends on:

- its mass
- what it is made of
- the temperature change.

1 The same mass of two different substances is heated. The amount of thermal energy transferred is the same in each case. Why does one material have a bigger increase in temperature?

Hot water

It takes a lot of energy to raise the temperature of 1 kg of water by a certain amount; more than for most other substances.

Specific heat capacity (c) is a measure of how much energy is required to raise the temperature of 1 kg of a substance by 1 °C.

You can calculate the amount of energy stored in a system when it is heated by using the equation:

change in thermal energy = mass × specific heat capacity × temperature change

$$\Delta E = mc\Delta \theta$$

where ΔE is in J, m is in kg, c is in J/kg °C and $\Delta \theta$ is in °C.

Example: Calculate the change in thermal energy when 2 kg of water is heated from 20 °C to 80 °C.

COMMON MISCONCEPTION
....
Heat and temperature are not the same thing. When something is heated, thermal energy is being transferred to it. Temperature is a measure of how hot or cold something is.

Substance	c in J/kg °C
Water	4200
Copper	380
Steel	450
Concrete	800

MATHS
....
In maths, the symbol Δ means the difference or change in a quantity. ΔE means a change in energy; $\Delta \theta$ means a change in temperature.

$\Delta E = mc\Delta\theta$

$= 2 \text{ kg} \times 4200 \text{ J/kg °C} \times (80 °C - 20 °C)$

$= 504\,000 \text{ J}$

$= 504 \text{ kJ}$

2 How much energy is needed to heat 1 kg of copper by 20 °C?

3 How much energy is given out when the temperature of 2 kg of steel falls by 30 °C?

Water has a very high specific heat capacity. This means it can absorb a large amount of thermal energy from a hot object for a given temperature change of the water. Water is used to cool many car engines. The thermal energy store in the engine decreases and the thermal energy store in the water increases. Energy is then transferred from the thermal energy store in the water to the store in the air surrounding the radiator.

Water can also release a lot of energy without a large temperature decrease. This makes it a very useful way of transferring large amounts of thermal energy around a house in a central heating system.

Figure 1.13 Water is used to cool many car engines

4 Explain why a hot-water bottle is so effective at warming a bed.

5 What is meant by specific heat capacity?

6 How much energy is needed to raise the temperature of 3 kg of steel by 15 °C?

More about specific heat capacity

Example: 0.5 kg of copper at 90 °C is added to 2 kg of water at 10 °C. Calculate the final temperature of the copper and water (θ).

Assume that the decrease in thermal energy store of the copper = increase in thermal energy store of the water.

Change in thermal energy store of copper

$= 0.5 \text{ kg} \times 380 \text{ J/kg °C} \times (90 °C - \theta) = 17\,100 - 190\theta$

Change in thermal energy store of water

$= 2 \text{ kg} \times 4200 \text{ J/kg °C} \times (\theta - 10 °C) = 8400\theta - 84\,000$

So $8400\theta - 84\,000 = 17\,100 - 190\theta$

$8590\theta = 101\,100$

$\theta = 101\,100/8590$

$\theta = 11.8 °C$

7 A night storage heater contains 50 kg of concrete. The concrete is heated during the night when electricity is cheaper, gradually emitting stored energy during the day. How much thermal energy is required to warm the concrete from 10 °C to 30 °C? Suggest why concrete is chosen.

8 A 1 kg steel block at 80 °C is added to 0.5 kg of water at 10 °C. Calculate the final temperature of the block and the water.

DID YOU KNOW?

Several tonnes of liquid sodium metal are used as a coolant in some types of nuclear reactor. The liquid sodium transfers thermal energy from the reactor core to water, which then turns to steam and drives the generators.

REQUIRED PRACTICAL

Investigating specific heat capacity

Learning objectives:

- use theories to develop a hypothesis
- evaluate a method and suggest improvements
- perform calculations to support conclusions.

A useful way of thinking about energy and energy transfers is the concept of stores. Energy can be stored in a variety of ways and one of those is by heating something up. If this object is put somewhere cooler then energy will be transferred from one store (the object) to another store (the surroundings).

Using scientific ideas to plan an investigation

When a lump of brass is immersed in a freezing mixture of ice and water its temperature decreases. It would end up at 0°C (as long as we left it in contact with the water for long enough). If we then take the brass out of the ice-cold water and put it into another beaker of hot water, thermal energy transfers from the store in the water to the brass. The brass warms up and the water cools down until they are both at the same temperature.

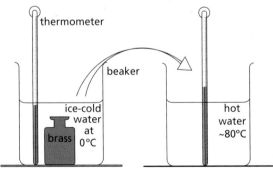

thermometer

beaker

ice-cold water at 0°C

brass

hot water ~80°C

Figure 1.14 The lump of brass is transferred from the water at 0 °C to the hot water

These pages are designed to help you think about aspects of the investigation rather than to guide you through it step by step.

DID YOU KNOW?

Stone age man, Native American Indians and backwoodsmen all have used hot stones to boil water. Hot stones from a fire are dropped into a wooden bowl of cold water. Thermal energy stored in the stones is transferred to the water, making it hot enough to boil. This is an example of a decrease in one **energy store** producing an increase in another. You can make use of this method to find the specific heat capacity of different materials.

KEY INFORMATION

~21 °C means about 21 °C

1. How could we find out what the temperature of the water became when the brass was added to it?

2. How could we calculate the decrease in temperature of the water?

3. How could we calculate the temperature rise of the brass?

4. What would happen to the temperature of the water and the brass in the second beaker if we left them for a long time (e.g. an hour)?

Evaluating the method

This experiment is used to find the **specific heat capacity** of brass by assuming that all the thermal **energy transferred** from the hot water increases the temperature of the lump of brass. We are equating the decrease in thermal **energy store** of the water to the increase in thermal energy store of the brass. If this assumption is not true, the method will not be valid.

5 The lump of brass has to be moved from one beaker to the other. Consider how this step in the method could affect the accuracy of the results.

6 The energy transferred from the water to the brass will cause the lump of brass in the second beaker to get hotter. Why will the energy transferred to the lump of brass not be stored there permanently?

7 What are the implications of your answers to questions 5 and 6 for the way the experiment is carried out?

8 Why is it important that the lump of brass is covered in water in the second beaker?

> **KEY INFORMATION**
>
> When thinking about this experiment, remember that energy tends to move from hotter areas to cooler ones.

Using data to calculate a value for specific heat capacity

We can *also* find the specific heat capacity of brass by transferring the brass from boiling water to cold water.

Decrease in thermal energy of brass = increase in thermal energy of water

$m_{brass} \times c_{brass} \times$ temperature decrease$_{brass}$

$= m_{water} \times c_{water} \times$ temperature increase$_{water}$

The final temperature of the water and brass is the same (they reach thermal equilibrium). As long as we know the values of the mass of water, mass of brass, specific heat capacity of water and initial temperatures of the water and brass, we can find the unknown value for the specific heat capacity of brass.

9 There is 250g of water (c_{water} = 4200 J/kg°C) in the second beaker and its temperature rises from 17°C to 26°C. Determine how much energy has been transferred into it.

10 How much energy can we assume has been transferred out of the brass when it is put into the second beaker?

11 If the brass had been in boiling water, by how much would its temperature have decreased?

12 The lump of brass has a mass of 600g. Calculate the specific heat capacity of brass.

13 Explain why is this method likely to give a lower value for the specific heat capacity than its true value.

Dissipation of energy

Learning objectives:

- explain ways of reducing unwanted energy transfer
- describe what affects the rate of cooling of a building
- understand that energy is dissipated.

KEY WORDS

conduction
energy dissipation
radiation
thermal conductivity

Thermograms (Figure 1.15) are infrared photographs in which colour is used to represent temperature.

Reducing energy transfer

Sometimes the transfer of thermal energy is useful, such as in cooking, but on other occasions we might want to reduce it. For example, pushing a supermarket trolley with stiff wheels needs a lot of work to be done against frictional forces. This means the wheels transfer some energy to the surroundings as thermal energy. This energy is wasted and we would want to reduce this energy transfer.

Lubrication – oiling the parts of a machine, or lubricating them, reduces the friction force. Less energy is wasted as thermal energy.

We might want to reduce thermal energy being transferred so that a parcel of fish and chips stays hot for longer or so that a block of ice cream stays frozen until we get it home.

Thermal insulation – surrounding a hot object with insulating material, or thermal insulation, reduces the rate at which energy is transferred away from it so the hot object cools more slowly (Figure 1.16). Clothing made of wool is a good insulator. Air is trapped between the wool fibres. Wool and air are bad **conductors** of thermal energy.

Figure 1.15 White represents the hottest area and blue the coolest

Figure 1.16 Thermal insulation in a roof

1 The wheels of a scooter do not move freely. Describe how unwanted energy transfers in the wheels can be reduced.

2 Explain why a wrapping of newspaper is as good at keeping fish and chips hot as it is at keeping a block of ice cream cold.

Insulating a building

A building needs to be well insulated so that less energy is needed to keep the building warm. Loft and cavity wall insulation reduce the rate of energy transfer from inside a building to the colder outside.

Look at Figure 1.17. White, red and yellow represent the hottest areas. Black, dark blue and purple represent the coldest areas.

DID YOU KNOW?

The foam blocks used for cavity wall insulation in new houses have shiny foil on both sides. This further reduces energy transfer, as shiny objects reflect **radiation** rather than absorbing or emitting it.

The diagram of thermal energy losses from a house (Figure 1.18) shows that it is important to insulate the walls and roof.

Insulation reduces the amount of energy transfer by **conduction**. The lower the **thermal conductivity** of the insulating material and the thicker the layer, the more the rate of energy transfer by conduction is reduced.

Figure 1.17 The wheels of the car are red because they are hot

3 Looking at Figure 1.18, where is most thermal energy lost? Explain how you could reduce this unwanted energy transfer.

4 Explain, using examples, how an eco home that has no heating system can stay warm.

Energy dissipation

A system is an object or group of objects. The total energy in a closed system is always constant. Energy is never created or destroyed.

Work done against the frictional forces acting on a moving object causes energy to be transferred from the object. This energy transfer raises the temperature of the surroundings by such a small amount that it is of no use. This is wasted energy – energy is transferred to a store where it cannot be used. Some energy is always **dissipated** when it is transferred.

5 Explain why we are unable to reclaim thermal energy arising from energy due to friction or air resistance.

6 A car engine is designed to transfer energy from fuel into kinetic energy, making the car move. Name the other energy transfers that are likely to take place.

7 Look at the thermal image in Figure 1.17. Describe some of the features you can see. Use technical terms such as *thermal energy*, *thermal conductivity* and *dissipated* in your description.

Figure 1.18 Thermal energy loss from a house

Energy efficiency

Learning objectives:

- explain what is meant by energy efficiency
- calculate the efficiency of energy transfers
- find out about conservation of energy.

Electric cars have an efficiency of about 85% compared with about 15% for a petrol car. This means an electric car can transfer a greater portion of energy in the form of kinetic energy from the original energy store. But batteries have to be charged – and the fossil fuel plant supplying the electric current is only about 35% efficient.

Useful energy output of a system

It is not possible for the useful energy output of a system to be greater than the total input energy. The law of conservation of energy also says that when a system changes, there is no change to the total energy of the system.

Even so, not all of the total energy input to a system is stored or usefully transferred. Some energy is always dissipated. This reduces the amount of energy that is usefully transferred.

For example, as an electric car accelerates, the engine transfers energy from the chemical energy store of the battery to the kinetic energy store of the car. Some of the input energy is wasted by transfer of thermal energy by heating the wheels and the surrounding air.

1 **What effect does the waste energy from a light bulb have on the surroundings?**

2 **Suggest why an electric vehicle is more efficient than a petrol or diesel vehicle.**

Calculating energy efficiency

Efficiency is an indication of how much of the energy supplied to a device is transferred as a useful output. If all of the energy supplied was transferred usefully the transfer would be 100% efficient.

$$\text{efficiency} = \frac{\text{useful output energy transfer}}{\text{total input energy transfer}}$$

OR

$$\text{efficiency} = \frac{\text{useful power output}}{\text{total power input}}$$

Figure 1.19 The least efficient light bulbs are being replaced with more energy-efficient alternatives, including compact fluorescent and LED light bulbs

MATHS

To convert from a decimal to a percentage, multiply the decimal by 100.

Quite a lot of energy transfers are not very efficient; a lot of energy is transferred in a way that isn't useful.

Example: Tina is an athlete. She applies a force of 75 N for a distance of 100 m using 44 000 J of chemical energy stored in food. What is her efficiency? Give your answer as a percentage.

Useful work done = force × distance = 75 N × 100 m = 7500 J

$$\text{Efficiency} = \frac{\text{useful energy output}}{\text{total energy input}}$$

$$= \frac{7500\,\text{J} \times 100}{44\,000\,\text{J}} = 17\%$$

When you exercise, you get hot – your body temperature increases. Most of the chemical energy stored in food is transferred to the thermal energy of your body.

3 For every 100 J of energy supplied to a motor, 80 J of useful work is done. Calculate the efficiency of the motor. Give your answer as a percentage.

4 For every 500 J of energy in coal, 135 J are transferred to a room as heat from a coal fire.

   a   Calculate its efficiency.

   b   Suggest why coal fires are inefficient.

5 Suggest why a kettle is not 100% efficient.

6 The efficiency of a television is 0.65. Calculate the useful energy output if the total energy input = 200 J.

Conservation of energy

Energy cannot be created or destroyed, only transferred from one store to another. In a closed system (one in which no energy can enter or leave) the *total* amount of energy put into the system equals the *total* amount of energy output. We say that energy is conserved. This is the law of **conservation of energy**. However, only some of the energy output is useful to us. The rest is dissipated as wasted energy. This affects the efficiency of a machine.

7 What is meant by 'conservation of energy'?

8 Explain why heating a material does not just increase the thermal energy store of the material.

9 When sound transfers energy from the store in a vibrating cymbal to your eardrums not all of the energy is transferred to the kinetic energy store of the eardrums. Suggest how some of the energy is wasted.

10 An electric car is 85% efficient. The electricity for the car is supplied by a coal power station with an efficiency of 35%. Determine how much energy is needed from the chemical energy store of the coal for the car to provide 100 J of useful energy.

DID YOU KNOW?

The efficiency of a filament bulb is less than 5% – over 95% of the energy supplied is transferred directly as thermal energy to the surroundings.

COMMON MISCONCEPTION

You often hear phrases like, '*Conserve energy*; turn off the lights'. However, to scientists, *conservation of energy* means that there is no net change to the total energy of a system.

REQUIRED PRACTICAL

Investigating ways of reducing the unwanted energy transfers in a system

Learning objectives:

- use scientific ideas to make predictions
- analyse data to identify trends
- evaluate an experimental procedure.

Energy will always be transferred from warmer places to cooler ones. Sometimes this is useful, as in a domestic heating system. However on other occasions we need to stop that transfer, or at least slow it down.

These pages are designed to help you think about aspects of the investigation rather than to guide you through it step by step.

Developing a hypothesis

Tazim's group are investigating **insulation** materials and they are trying to predict which will work well. They are going to use the material to pack around a hot cooking pot that will be put into a wooden box with a close-fitting lid. The pot also has a lid. The liquid in the pot needs to be kept as hot as possible.

They are investigating the use of:

- expanded polystyrene
- wood shavings
- air.

One of the ways that thermal energy is transferred is by **conduction**.

1 Tazim thinks that all the materials will have the same insulating effect, because only metals are good thermal conductors. Is he right?

2 Write a list of all the factors that may affect the thermal insulation properties of the box around the hot cooking pot.

3 Write a hypothesis connecting one of these factors to the temperature of the liquid in the pot.

4 Plan a method that would enable the hypothesis to be tested. Include in this an indication of the variables to be kept the same, the variables to change and the variables to measure.

Analysing the results

The students then set up an experiment to investigate the cooling of the cooking pot when insulated with different materials. They left a similar pot on the bench to act as a control. The results are shown in the table.

Thermal insulator	Temperature of the pot (°C)					
	Start	1 hour	2 hour	3 hour	4 hour	5 hour
Control	90.0	38.0	26.5	25.5	25.0	25.0
Air	90.0	43.5	28.0	25.5	25.5	25.0
Polystyrene	90.0	55.0	43.5	37.5	30.0	26.5
Wood shavings	90.0	52.5	40.0	32.5	28.0	25.5

5 Draw a graph of how the temperature of the pot changes with time for each insulator.

6 What feature on the graph indicates which material is best at reducing energy transfer?

7 Explain why the graph lines levelled out.

8 Suggest how you could calculate the rate of cooling from the graphs.

9 Suggest why the rate of cooling changed from during the first hour to during the fifth hour.

Evaluating the experiment

Think about their experiment and the evidence it produced. The students are sharing ideas about how effective the experiment was.

10 These are some of the comments made. Respond to each of these, justifying your point of view:

a 'This experiment didn't work because all the pots ended up nearly cold.'

b 'To make it a fair test we should have used the same mass of liquid in each experiment.'

c 'The control was pointless – it's just using air as an insulator.'

d 'If you draw a straight line from the starting temperature to the final temperature for each of the experiments they have nearly the same gradient, showing that there's little difference between the types of insulation.'

11 a Show that the energy transferred to the surroundings in the first hour was about 1.33 times more when air was used as the insulator compared with polystyrene.

b Calculate what this value would be for the second hour.

DID YOU KNOW?

Figure 1.20 A haybox

During the Second World War, fuel for cooking was rationed and people were encouraged to use an older method of cooking, using a haybox. Hayboxes were ideal for slow cooking simple foods such as stews and soups. Food would be heated up using conventional methods and then put in the haybox to finish cooking. A haybox works by greatly reducing the transfer of energy from the hot food to the surroundings. As the name suggests, they were boxes packed with hay or shredded paper. This traps still air which acts as an excellent insulator.

ADVICE

Measuring the rate of cooling in the first hour gives a good basis for comparing the materials.

Using energy resources

Learning objectives:

- describe the main energy sources available for use on Earth
- distinguish between renewable and non-renewable sources
- explain the ways in which the energy resources are used.

Across the world we use over five hundred million trillion joules of energy every year. This has to come from somewhere.

The need for energy

Industrial societies use huge amounts of energy. Much of it is used for the generation of electricity in power stations. Electricity is useful because appliances can use it to transfer energy to many different types of useful energy stores. For example, a kettle can use electricity to transfer energy to the thermal energy store of the water that it is heating up.

1 **Identify an electrical device that transfers energy to:**

   a **a kinetic energy store**

   b **a gravitational potential energy store.**

We can also use electricity for transport (such as trains and electric cars) and for heating. However, it is often cheaper and easier to use energy resources directly for these purposes.

2 **Give an example of an energy resource being used to heat something.**

3 **State the energy resource you are using when you ride a bicycle.**

4 **Suggest why we mainly use electricity to power trains but fuel to power aeroplanes.**

Figure 1.21 A gravity light has been developed that transfers the gravitational potential energy of a falling 12 kg weight to power a light

DID YOU KNOW?

Tidal energy comes from the gravitational potential energy store of the Moon. Nuclear and geothermal resources store energy that was produced by stars exploding in a supernova.

Energy resources

Most of the energy resources available to us store energy that has originally come from the Sun. These include fossil fuels (coal, oil and gas), biofuel, wind, wave, hydroelectric and solar energy. Other resources include geothermal energy (from hot, underground rocks), tidal energy and nuclear energy.

All of these resources have advantages and disadvantages. For example, in fossil fuels the energy is very concentrated – a small amount of fuel stores a large amount of chemical energy. Biofuels (such as wood pellets) are produced from plants and animals. They are not as concentrated as fossil fuels so you need a much larger mass to release the same amount of energy.

Reliability is another issue. Some energy resources are available all the time but others are not. It is difficult to capture solar energy during bad weather or at nighttime. Rain is needed for hydroelectric power, and wind and wave energy require it to be windy.

5 Give two ways of generating electricity in which no fuel is burned.

6 Suggest why wave energy is only useful for a few countries.

7 Satellites orbiting the Earth use solar panels to provide them with their energy needs. Suggest why they also carry rechargeable batteries.

Figure 1.22 Wind turbines are usually placed on high hills or out at sea

Renewable and non-renewable energy resources

Renewable resources never run out (at least not for the foreseeable future). They can be replenished as soon as they are used.

- Solar, wind, wave and hydroelectric resources are intermittently available and the quantity varies according to location. The Sun always shines and it always creates wind and rain. However, there may be short periods of time when there is little wind or rain and no waves.

- Any thermal energy extracted from hot rocks from the geothermal resource is replenished by physical processes in the rocks.

- The Moon produces tides every day, whether we extract energy from them or not.

8 Wood is a form of biofuel. Explain why we can regard wood as a renewable resource.

Non-renewable resources will run out eventually. This is because they are not being replenished at the same rate that we are using them.

9 Identify a non-renewable resource that is not a fossil fuel.

10 Both biofuels and fossil fuels are formed from plants and animals. Explain why biofuels are renewable resources but fossil fuels are non-renewable.

11 a Describe the difference between energy and energy resource.

   b Explain why energy is conserved but an energy resource is not conserved.

Global energy supplies

KEY WORD
....
efficiency

Learning objectives:

- analyse global trends in energy use
- understand what the issues are when using energy resources.

. . . .

The white clouds coming from a power station are not smoke but water vapour coming from the cooling towers. In a coal-fired power station about half the energy stored in coal is lost in this way. We need the electricity, but do we have to have the waste?

Global trends

Figure 1.23 shows how energy resources have been used in the past and how they are likely to be used in the future.

You can see that the amount of energy we need increases every year. This is partly because the population of the world is increasing. Also, more countries are developing technologically and need more energy for industry. China's energy use, for example, grew rapidly between 2000 and 2015.

Although there are environmental problems with fossil fuels, they are likely to provide the world with most of its energy for many years to come.

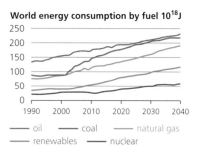

World energy consumption by fuel 10^{18}J

—— oil   —— coal   —— natural gas
—— renewables   —— nuclear

Figure 1.23 The world's use of energy resources

1. **Explain why some of the lines on the graph in Figure 1.23 will eventually stop rising.**

2. **Look at Figure 1.23. Which single type of fuel is used the most?**

3. **Suggest why fossil fuels are likely to provide most of the world's energy for the foreseeable future.**

Energy issues

Scientists have discovered that using some energy resources is having an environmental impact. For example, fuel-burning power stations cause pollution.

They release carbon dioxide into the atmosphere which traps the Sun's energy and contributes to global warming. They also emit sulfur dioxide which causes acid rain.

Nuclear power stations create waste, some of which remains dangerous for thousands of years. There is also the danger of a nuclear accident which could contaminate the area with radioactive materials.

Scientists do not always have the power to make the decisions about energy use – these decisions are made by governments.

DID YOU KNOW?
....

Mean sea level has risen by about 20 cm in the last 100 years. Most scientists think this is due to global warming. Levels are expected to rise as much as a further 240 cm in the next 100 years.

Governments have to weigh up ethical, social and economic considerations as well as the likely impact on the environment. They also need to make sure they represent the views of the population that elected them.

Ethical considerations are to do with whether something is morally right or wrong. An example is making sure that energy resources are plentiful for the future. Although it will not affect us, it would be wrong to cause problems for future generations.

Social considerations concern how they affect people. Fossil fuel power stations need to be built to power factories and so provide people with jobs.

Economic considerations involve money. A particular technology might be less harmful to the environment but be far more expensive to construct or maintain.

Figure 1.24 A protest sign against plans to extract shale gas from the ground

4 A local council has plans to extract shale gas from the ground and to burn it as fuel. Suggest some considerations that the council needs to make before they decide whether to go ahead.

HIGHER TIER ONLY

Using energy resources efficiently

One way of preserving energy resources is to use them efficiently. This means you need to find ways to reduce the amount of wasted energy in any energy transfer.

The table shows the efficiencies of some typical engines and motors.

	Efficiency (%)
Petrol engine	25
Diesel engine	35
Electric motor	80

Fuel-burning engines have a low efficiency. It is impossible to transfer thermal energy into kinetic energy without wasting most of it.

You can use this equation to calculate **efficiency**:

$$\text{Efficiency} = \frac{\text{useful power output} \times 100}{\text{total power input}}$$

5 Suggest why the electric motor is more efficient than the fuel-burning engines.

6 The output power of an electric motor is 3 kW. Calculate the input power it needs.

7 The power station that produced the electricity for the electric motor is 35% efficient. Use this value and your answer to question 6 to determine whether using an electric motor is more efficient than a diesel engine.

8 Besides increasing the efficiency of the engine, what else could you do to increase the efficiency of a car?

Energy transfer

Learning objectives:

- understand why energy is a key concept in science
- use ideas about stores and transfers to explain what energy does
- understand why accounting for energy transfers is a useful idea.

KEY WORDS

chemical
dissipate
energy
store
thermal
transfer

Young children often love to run around and play on swings. After a while they'll get tired. 'I've run out of energy!' they'll say, and flop down. A few minutes later they're up again, announcing 'I've got my energy back!' and run off again.

What is energy?

The concept of **energy** is one of the most important ideas in science. We use it to explain what's happening when a torch is turned on – and when the batteries run flat. We talk about an energy crisis, and whether we have enough energy to keep our lamps lit and our cars on the road. It's important in chemistry because when reactions take place energy is transferred, and it's important in biology because cells need energy to carry out important processes.

However, we sometimes find it difficult to explain what we mean by energy. It's easier to understand what stores energy, rather than what energy is.

1 **Decide which of these is an energy store:**

    a a ball rolling down a ramp

    b a stretched string

    c a hot object

    d a mixture of oxygen and fuel.

2 **For each of a to d in Question 1, suggest if and how energy might be transferred from one store to another and used for something useful.**

Transferring energy

Sometimes we use names for energy, such as electrical, kinetic or gravitational potential. Actually it's more useful to identify where the energy is. This tells us more. We can then work out how it's been transferred.

Think about a wind-up torch (Figure 1.25). When you turn the handle you transfer energy from your muscles to the

Figure 1.25 Energy is transferred into this torch when it is wound up and out of it when it used to project light

energy store in the torch's battery. The energy store in your body decreases while the the energy store in the battery increases. When you turn the torch on, the energy store in the battery decreases and an electric current transfers energy in the bulb. The internal energy of the bulb increases (it increases in temperature), and the bulb transfers energy by light waves to the surroundings. The bulb **dissipates** energy to the surroundings, increasing the internal energy of the surroundings. When the battery is flat, its energy store is zero and all the energy has been transferred to the surroundings.

3 Draw a flow diagram to represent where energy has been transferred from and to.

4 Identify which energy store increases and which decreases:

  a in a bicycle, being ridden

  b in a match, being struck.

Accounting for energy

Energy does not disappear when it is transferred to other stores. It is still there but it is sometimes hard to see where it is being stored. For example, a charged battery is storing energy. When the battery is flat, the energy has been transferred to the surroundings.

Figure 1.26 Accounting for energy is rather like checking on the number of bricks in a set

The physicist Richard Feynman suggested that one of the ways of understanding accounting for energy was to compare it to a set of child's building blocks. The parent knows how many blocks there are in the set. If there is a block missing when you put them away, you look for it. Energy is similar. When you start with a certain amount of energy in a system, it is all still there, but you may have to look for it.

Coal has a **chemical** energy store. When the coal is burnt, the chemical energy store of the coal decreases and the thermal energy store of the coal increases. In a steam engine, energy is transferred from the store in the coal to the water (as steam) in the boiler. The steam turns the wheels and makes the train move, so there is a kinetic energy store associated with the train. At the end of the journey the coal is burnt, the steam used and the train has come to rest, but the energy is still there – the thermal energy store of the surroundings has increased.

Figure 1.27 Energy is being transferred here; can it all be accounted for?

5 How does the story of the child's bricks explain the concept of conservation of energy?

6 Energy is conserved within a system. In the case of the steam engine, what does the system include?

Because scientists think this is an important idea, they go to a lot of trouble to measure the amounts of energy (look at the labels on prepared food), to account for the energy transfers and to calculate the efficiency from this (which can be given as a percentage).

MATHS SKILLS

Calculations using significant figures

Learning objectives:

- substitute numerical values into equations and use appropriate units
- change the subject of an equation
- give an answer to an appropriate number of significant figures.

KEY WORDS

rearrange an
    equation
significant figures
subject of an
    equation
substitute

People have been trying to invent a perpetual motion machine for centuries. Perpetual motion means moving forever. But where there are energy transfers, some energy is always dissipated. This is why the idea of perpetual motion is an impossible dream.

Calculating changes in energy

Example: A diver dives into water from a board 10.0 m above the water surface. The mass of the diver is 50.0 kg. The gravitational field strength is 9.8 N/kg.

a Calculate the change in gravitational potential energy of the diver as she falls from the board to the surface of the water.

b Calculate the diver's speed when she hits the water.

a Use $E_p = mgh$

$E_p = 50.0 \text{ kg} \times 9.8 \text{ N/kg} \times 10.0 \text{ m}$

$= 4900 \text{ J}$

Figure 1.28 An idea for a perpetual motion machine. Such a machine could not work because it would violate the law of conservation of energy.

b As the diver falls, energy is transferred from a gravitational potential energy store to a kinetic energy store.
The energy transferred from the gravitational potential energy store to the kinetic energy store = 4900 J

Use $E_k = \frac{1}{2}mv^2$

You need to **rearrange the equation** to find v (make v the **subject of the equation**).

Multiply both sides by 2: $2E_k = mv^2$

Divide both sides by m: $\dfrac{2E_k}{m} = v^2$

So: $v = \sqrt{\dfrac{2E_k}{m}}$

Substituting the values for E_k and m:

$v = \sqrt{\dfrac{2 \times 4900}{50}}$

$v = \sqrt{196}$

$v = 14 \text{ m/s}$

1 Calculate the change in gravitational potential energy of a diver of mass 70.0 kg who dives from a board 2.0 m above the water.

2 Calculate the gravitational potential energy gained by a ball of mass 50.0 g thrown to a height of 10.0 m.

Significant figures

The answer to a calculation can only have the same number of **significant figures** as the data provided.

Example: A diver stands on a board 5.0 m above water. The mass of the diver is 50.0 kg. The gravitational field strength is 9.8 N/kg. Calculate the diver's speed when he hits the water. Give your answer to two significant figures.

Increase in kinetic energy store = decrease in gravitational potential energy store

$= mgh$

$= 50.0 \text{ kg} \times 9.8 \text{ N/kg} \times 5.0 \text{ m}$

$= 2450 \text{ J}$ (Do not round to two significant figures yet.)

Energy transferred to the kinetic energy store, $E_k = 0.5 \, mv^2$

Rearranging as before, $\dfrac{2E_k}{m} = v^2$

$v = \sqrt{\dfrac{2E_k}{m}}$

$= \sqrt{\dfrac{2 \times 2450}{50}}$

$= \sqrt{98}$

$= 9.899$, or 9.9 m/s, to two significant figures

3 A spring of spring constant $k = 350$ N/m is extended by 9.0 cm. Calculate the elastic potential energy, E_e, stored in the spring. Give your answer to two significant figures.

Two-step problems

Example: A lift raises four people with a combined weight of 3500 N to the third floor of a building in 25.0 seconds. The height gained is 15 m. Calculate the power required.

You need to use: power $= \dfrac{\text{work done}}{\text{time taken}}$, or $P = \dfrac{W}{t}$

You have the time taken, but not the work done. You must approach this problem in two steps. First, calculate the work done, then **substitute** that value into the equation for power.

Work done = force × distance moved along the line of action of the force

The force used (the weight that is lifted) is 3500 N.

Work done = 3500 N × 15 m

$= 52\,500$ J, or 52.5 kJ

Substituting this value for work done into the power equation:

power $= \dfrac{52\,500 \text{ J}}{25.0 \text{ s}}$

$= 2100$ W, or 2.1 kW

KEY INFORMATION

You should keep at least one extra significant figure in your calculations to avoid rounding errors. Only round to the required number of significant figures when you have calculated the answer.

KEY INFORMATION

To round a number to two significant figures, look at the third digit. Round up if the digit is 5 or more, and round down if the digit is 4 or less.

MAKING LINKS

A set of readings taken with the same instrument should all have the same number of significant figures. So if the extension of a spring is measured with a ruler as 4.1, 4.2, 5.9 and 6 cm, the table of data should list 6 cm as 6.0 cm.

MATHS SKILLS

Handling data

Learning objectives:

- recognise the difference between mean, mode and median
- explain the use of tables and frequency tables
- explain when to use scatter diagrams, bar charts and histograms.

KEY WORDS

anomalous   independent
bar chart   variable
continuous data   line of best fit
correlation   mean
dependent   median
   variable   mode
frequency table   scatter diagram
histogram

The purpose of an experiment is to find out the relationship, if there is any, between the variables you are investigating. You do this by looking for a pattern in the data that is collected. It is difficult to spot a pattern from a mass of numbers, but it is much easier from a picture – a graph or a chart.

Tables and frequency tables

Tables are used to capture the information from an experiment and from graphs. A simple table is used for **continuous data**, where information is measured on a continuous scale – for example, when loading a spring with different weights and measuring the extension produced (Table 1.1).

Force (N)	Extension of spring (mm)		
	1st reading	2nd reading	3rd reading
10	8	9	7
20	19	16	15
30	23	21	29
40	34	32	33
50	35	44	39

Table 1.1 Investigating the relationship between force and the extension of a spring

Another way of collecting data is to use tally marks and make a **frequency table**. This method is best where your data can only have certain values such as shoe sizes, or the information can be categorised such as the names of countries.

Using scatter diagrams, bar charts and histograms

The data in Table 1.1 is continuous and is best displayed by using a **scatter diagram** (Figure 1.29). The force is the **independent variable** so it goes on the horizontal axis. The extension is the **dependent variable** so it goes on the vertical

Country	Tally marks	Number of new nuclear power stations being built
Russia	IIII II	7
UK	IIII	5
India	IIII I	6
USA	IIII	5
Pakistan	III	3

Table 1.2 Number of new nuclear power stations being built in selected countries

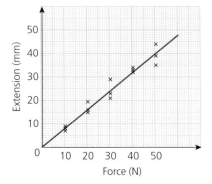

Figure 1.29 Data from Table 1.1: A scatter graph with a line of best fit

axis. Plotting the data points will give a clearer picture of the relationship between the two variables, to see if there is a **correlation**. If there is a correlation, you can draw a **line of best fit** (this may be a curve) and then use this to extrapolate further trends and information.

The data in Table 1.2 is not continuous. It is best represented by a bar chart (Figure 1.30).

The data in Table 1.3 is grouped. It is best represented by a **histogram**. Each column represents a group of data, and the frequency of the data is shown by the area of each bar (Figure 1.31).

Age	Tally	Frequency
11–20	ЖІ ІІІІ	9
21–30	ІІІІ⁄ІІІІ⁄ІІІІ⁄І	16
31–40	ЖІ ЖІ ІІ	12
41–50	ІІІІ	4
51–60	ЖІ І	6
61–70	І	1

Table 1.3 Frequency distribution of the ages of people on a street

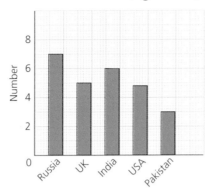

Figure 1.30 Data from Table 1.2: A bar chart

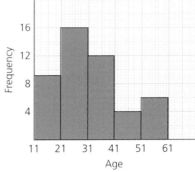

Figure 1.31 Data from Table 1.3

Mean, median and mode

To find the **mean**, add up all the values and divide by the number of values.

Example: The mean of four repeated measurements of temperature, 7 °C, 12 °C, 9 °C and 11 °C, is

$$\frac{7\,°C + 12\,°C + 9\,°C + 11\,°C}{4} = 9.75\,°C$$

or 10 °C, rounded to the nearest whole number.

The **median** is the middle value when the values are rearranged in sequence. For example, from the data set above:

7 °C, 9 °C, 11 °C, 12 °C

When there are an odd number of data points, it is easy to spot the median. If there is an even number of data points, calculate the median as the midpoint of the two middle values. In this example, the median is halfway between 9 °C and 11 °C. So the median is 10 °C, even though that is not one of the data points.

The **mode** is the value, or item, that occurs most frequently in a set of data. It is easy to spot the mode in a frequency table. In Table 1.4 the mode is 'Biomass & renewable waste'.

When should you use which average? In the repeated measurements of temperature above, the mean and the median turn out to be the same value. But if one of the measurements was **anomalous**, for example, 22 °C instead of 12 °C, this would skew the mean.

Renewable energy resource with the greatest share of total renewable energy production	Frequency
Solar energy	2
Biomass & renewable waste	31
Geothermal energy	0
Hydropower	0
Wind energy	1

Table 1.4 Renewable energy sources in the EU for 2013

Example: The mean of 7 °C, 22 °C, 9 °C and 11 °C is

$$\frac{7\,°C + 22\,°C + 9\,°C + 11\,°C}{4}$$

= 12.25 °C

or 12 °C, rounded to the nearest whole number.

The median, 10 °C, is unchanged, so it is the more accurate representation of your data set when there are some anomalous results (or 'outliers').

Check your progress

You should be able to:

Describe how energy can be stored by raising an object up or by stretching or compressing it → Use the equations for gravitational potential energy and elastic potential energy → Apply the equations for gravitational potential energy and elastic potential energy in a variety of contexts, and change the subject of these equations

Describe how a moving object has kinetic energy → Know that kinetic energy is related to mass and velocity squared and use the equation to calculate it → Use the equation for kinetic energy to solve problems, including changing the subject of the equation

Recognise that when a force moves an object along the line of action of the force, work is being done

State that various devices do work and, in doing so, transfer energy → Calculate the work done by a force from the size of the force and the distance moved → Use the equation for work done to solve problems, including changing the subject of the equation

State that some materials require more energy than others to increase a certain mass by a certain temperature rise → Describe what is meant by the specific heat capacity of a material and use the equation for specific heat capacity

Plan an experiment to measure the specific heat capacity of a material → Calculate temperature changes, masses or specific heat capacities given the other values

Evaluate an experiment to measure the specific heat capacity of a material

Recognise that some energy transfers are unwanted → Describe how lubrication and insulation can be used to reduce unwanted energy transfers

Calculate energy efficiency

Describe how some energy transfers are more useful than others → Explain how thermal conductivity affects the rate of energy transfer across a material and affects the rate of cooling of a building

Recognise that in a closed system there may be energy transfers that change the way energy is stored, but there is no net change to the total energy

State that various resources are used as fuels and to generate electricity → Describe the advantages and disadvantages of fossil fuel, nuclear and renewable energy resources → Evaluate and justify the use of various energy resources for different applications

Worked example

Jo's group is investigating the energy changes that take place when a toy car runs down a ramp.

1 The mass of the car is 500 g, g is 9.8 N/kg and the vertical height of the top of the ramp is 20 cm. Calculate the GPE of the car at the top of the ramp.

$GPE = mgh = 500 \times 9.8 \times 20 = 98\,000\,J$

> This answer correctly uses the equation to multiply the variables together but hasn't converted the mass or the height into standard units. It should be 0.5 kg × 9.8 N/kg × 0.2 m = 0.98 J

2 The pupils use a speed detector to measure the speed of the car when it gets to the bottom of the ramp. They find out it is travelling at 0.8 m/s. Calculate its kinetic energy at the bottom of the ramp.

$KE = \frac{1}{2}mv^2 = \frac{1}{2} \times 0.5 \times 0.8 \times 0.8 = 0.16$

> This answer correctly uses the equation and has converted g to kg. The speed (only) is squared but there is no indication of the correct unit at the end.

3 a How could they calculate the percentage efficiency of the system at converting GPE into KE?

b Explain why it will be less than 100%.

Efficiency could be found by dividing output by input,

so $\frac{KE}{GPE}$.

Nothing is perfect and efficiency is always less than 100%

> This is correct in that it is the output divided by the input but it is important to use the **useful** output. Furthermore, the question asked for the efficiency as a percentage, so the answer has to be multiplied by 100.

> This is true but is not an explanation. The reason it is not 100% is because not all of the GPE store has been transferred to the kinetic energy store of the car. Some energy is dissipated to the surroundings, raising the temperature of the car body and the surrounding air.

4 Jo says that if the ramp was 100% efficient then doubling the height of the top of the ramp would double the speed at the bottom. Referring to the relevant formulae, suggest whether she is correct.

$GPE = mgh$ and $KE = 1/2mv^2$. Increasing h will increase GPE and this will increase KE which will increase v so Jo is correct.

> This quotes the correct formulae and identifies that an increased h will mean an increased GPE. In fact, doubling h will double the GPE which will double the KE (if it's 100% efficient). However, $KE = \frac{1}{2}mv^2$ and the squaring means that although the speed will increase, it won't double (it will actually go up by a factor of about 1.4). If the question includes a quantitative approach (it refers to doubling) it's not enough to give a qualitative response (say **how much** it increases).

End of chapter questions

Getting started

1. Sally is choosing a new electric kettle. One is rated at 1.5 kW and the other at 2 kW. What does this show? `1 Mark`

   a   The first one is smaller.

   b   The second one is a newer design.

   c   The first one will keep the water hot for longer.

   d   The second one will transfer energy more quickly.

2. Write down the equation for efficiency. `1 Mark`

3. What is meant by a non-renewable resource? `1 Mark`

4. When energy is being wasted we say it is:

   a   diffusing

   b   propagating

   c   dissipating

   d   refracting `1 Mark`

5. Describe the difference between elastic potential energy and gravitational potential energy. `2 Marks`

6. Two steel blocks, one with a mass of 100 g and the other with a mass of 200 g, are placed in boiling water for several minutes. Which of these statements is **not** true? `1 Mark`

   a   They are both at the same temperature.

   b   They are made of the same material.

   c   They have the same amount of stored thermal energy.

   d   They remain solid.

7. Order these light bulbs from most to least powerful: `1 Mark`

   15 W, 0.1 kW, 60 W, 0.08 kW, 150 W

8. Elise has a weight of 400 N and can clear a high jump of 2 m. Calculate how much work she does in raising her body up 2 m. `2 Marks`

Going further

9. Which of these does **not** affect the amount of energy needed to heat up a sample of material? `1 Mark`

   a   Its colour

   b   Its mass

   c   Its specific heat capacity

   d   The temperature rise

10. Explain why bubble wrap is an effective insulator. `2 Marks`

11. A hot parcel of fish and chips is taken outside on a cold night. Describe what movement of energy will take place, identifying where the energy is stored. `2 Marks`

12 Look at the image of the house and suggest what the thermogram shows.

13 A cat falls from a tree onto the ground. Describe the energy transfers that takes place. 3 Marks

More challenging

14 The power of a kettle is 2000 W. Explain what **2000 W** means in terms of energy transfer. 1 Mark

15 Write down what each of the symbols stands for in $\Delta E = mc\Delta\theta$ 1 Mark

16 Night storage heaters heat up during the night time and release energy during the day. Explain why they are made from a material with a high specific heat capacity. 2 Marks

17 A student was given some rods made from different materials. He was asked to arrange the rods in order of increasing thermal conductivity. Describe a procedure the student could follow to carry out the task safely by using a fair test. 6 Marks

Most demanding

18 Explain the difference between thermal energy and temperature. 2 Marks

19 Explain why energy is conserved within a closed system. 2 Marks

20 A student was investigating how the efficiency of a squash ball changed with temperature. The efficiency of a squash ball can be worked out by using the equation:

gravitational potential energy associated with the squash ball when it reaches its maximum height after it bounces/gravitational potential energy associated with the squash ball before it is dropped.

The student placed the ball in a beaker of water and heated the water to the desired temperature. She then removed the ball from the water and dropped it from a height of 1 m onto a hard surface. Once the ball had bounced, the student measured the height that the ball bounced up to by using a metre ruler.

Here are her results:

Mass of squash ball = 25 g

Gravitational field strength on Earth = 9.8 N/kg

Temperature (°C)	20	30	40	50	60
Bounce height 1 (cm)	6	28	39	43	45
Bounce height 2 (cm)	18	23	36	40	42

The student concluded that the higher the temperature of the squash ball, the more efficient it was.

Discuss whether the student made a valid conclusion and evaluate the limitations of the experiment. 6 Marks

Total: 40 Marks

ELECTRICITY

STATIC ELECTRICITY

- Electrical insulators can be charged by rubbing (friction).
- There are two kinds of electric charge – positive and negative.
- Like charges repel, unlike charges attract.

ELECTRIC CURRENT

- An electric current is due to a flow of charge.
- Resistors are used to control the current in circuits.
- Ohm's law is used to calculate resistance.

ELECTRICAL ENERGY

- Cells are sources of electricity.
- A battery is a number of cells joined together.
- Energy sources can be used to drive electrical generators to produce electricity.

ELECTRICITY IN THE HOME

- Electricity supplied to the home is alternating current (a.c.).
- Care must be taken when using mains electricity as it has a high potential difference (230 V).
- Fuses and circuit-breakers switch off the current if a fault occurs.

IN THIS CHAPTER YOU WILL FIND OUT ABOUT:

WHAT IS STATIC ELECTRICITY?

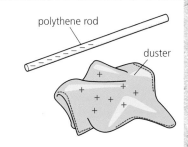

polythene rod

duster

- Insulators become charged when they gain or lose electrons.
- When two materials are electrically charged, there is an electric field between them.
- Static electricity can give a person an electric shock and it may create a spark which could cause petrol vapour or natural gas to explode.

WHAT ARE THE KEY CONCEPTS IN ELECTRICITY?

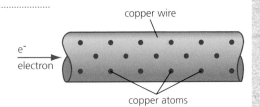

copper wire

e^-
electron

copper atoms

- Electric current is the rate of flow of charge through a conductor.
- Some electrical components resist the flow of electrons more than others so have greater resistance.
- Potential difference is a measure of the energy transferred per unit charge as charges move between two points in a circuit.

WHAT ARE THE CHARACTERISTICS OF SOME ELECTRICAL COMPONENTS?

- When electrical components are connected in parallel, there is more current passing through each component than when they are connected in series.
- A fixed resistor at constant temperature obeys Ohm's law, so the current through it is directly proportional to the potential difference across it.
- Diodes, thermistors and light-dependent resistors do not obey Ohm's law.

HOW CAN ELECTRICITY BE USED SAFELY IN THE HOME?

- Circuit-breakers and fuses are used to cut off the current when there is a short circuit.
- The earth is connected to the Earth so it provides a path for the current if there is a short circuit.
- The higher the power (in W) of an electrical device, the more expensive it is to use it.

Static electricity

Learning objectives:

- describe how insulating materials can become charged
- know that there are two kinds of electric charge
- explain these observations in terms of electron transfer.

KEY WORDS

attract
conductor
electron
insulator
repel

Explosions can be caused by a spark from the discharge of static electricity.

Producing static electricity

Clouds become charged as small pieces of ice bump into each other. If enough charge builds up, a spark in the form of lightning jumps across the gap between the cloud and the ground and an electric current passes between the cloud and Earth (Figure 2.1) or between clouds.

Figure 2.1 Cloud-to-ground lightning

Metals are good electrical **conductors.** Electric charges can move through them. But electric **insulators** such as glass, polythene or wood do not allow electric charges to move through them. Charge builds up. Many insulators can be charged by friction.

When a polythene rod is rubbed with a duster it becomes charged and can **attract** tiny pieces of paper (Figure 2.2).

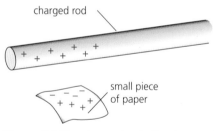

charged rod

small piece of paper

Figure 2.2 The charged rod attracts a small piece of paper by inducing a charge on the paper

Other materials can be charged by friction:

- When a balloon is rubbed on a sweater it becomes charged and can stick to a wall.
- Some dusting brushes are designed to become charged and attract dust.
- When hair is combed with a plastic comb both the comb and the hair become charged.

1 Name:

   a an insulator and

   b a conductor of electricity.

2 Explain why static charges do not build up on a conductor.

Two kinds of electric charge

There are two kinds of electric charge, positive and negative (Figure 2.3).

Like charges **repel**, unlike charges attract.

Forces of attraction and repulsion between charged objects are examples of non-contact forces.

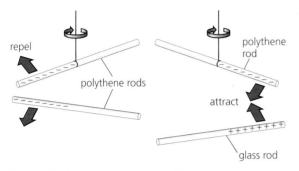

Figure 2.3 Like charges repel, unlike charges attract

3 **Polythene, acetate and Perspex rods are charged with a duster. State whether these pairs of rods will attract or repel each other:**

   a **a polythene rod and an acetate rod**

   b **two Perspex rods**

   c **a Perspex rod and a polythene rod**

   d **an acetate rod and a Perspex rod.**

4 **A Van de Graaff generator produces a large electrostatic charge. Beth puts her hands on a Van de Graaff generator when it is switched off. When it is switched on she becomes charged (Figure 2.4). Explain why does her hair stand on end.**

Figure 2.4 A Van de Graaff generator

Electron transfer

An atom consists of a small positively charged nucleus surrounded by negatively charged **electrons**. In a neutral atom there are equal numbers of positive and negative charges. All electrostatic effects are caused by the movement of electrons.

- If a polythene rod is rubbed with a duster, electrons **move from the duster to the polythene** making the polythene negatively charged.
- If an acetate rod is rubbed with a duster, electrons move **from the acetate to the duster** making the acetate rod positively charged.

5 **When a polythene rod is charged by rubbing it with a duster what charge, if any, does the duster gain?**

Electric fields

Learning objectives:

- explain what an electric field is
- draw an electric field pattern for a charged sphere
- use the idea of an electric field to explain electrostatic attraction and sparking.

KEY WORDS

electric field
spark

The electric field created by a charged object stretches out into space for ever. That means it can attract or repel another charge right across the universe. The force would be very weak though.

What is an electric field?

Charged objects can attract and repel each other because they create electric fields around themselves. Figure 2.5 shows the electric field that is created by a positive charge. If you place a second charge within this field, then it will feel a force of attraction or repulsion from the charge that has created the field.

A field is a region of space where a force can act at a distance. This means that objects do not need to be in contact with each other to exert a force – they just need to be within the other object's field. An **electric field** is a region where a force acts on a charged particle.

> **KEY INFORMATION**
>
> There are other types of fields as well as electric fields. These include gravitational and magnetic fields.

> **KEY INFORMATION**
>
> An electric field has a direction. In Figure 2.5 this is shown by the arrows on the field lines. The direction of an electric field is always the direction that a force would act on a positive charge inside the field.

1 Explain why all of the arrows are pointing away from the charge in Figure 2.5.

2 Describe how the electric field in Figure 2.5 changes when the positive charge is replaced with a negative charge.

The strength of an electric field

The electric field in Figure 2.5 gets weaker as the distance from the charge increases. A charge that is close to the centre will feel a stronger force of attraction or repulsion than one that is further away.

3 An electron moves in a circle around the charge in Figure 2.5. The charge is at the centre of the circle. State what happens to the size of the electric force on the electron.

4 Copy Figure 2.6. Label it to show a place where the strength of the field is:

a strong

b weak.

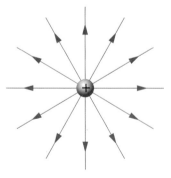

Figure 2.5 The electric field round a point or spherical charge

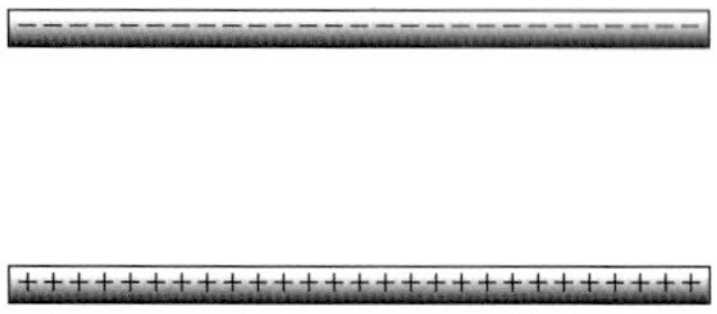

Figure 2.6

5 An electric field is represented by a set of lines with arrows on. Suggest why this is a useful way of showing a field and a way in which it might be misleading.

Sparking

The atoms and molecules of gases in the air contain positive and negative charges. Normally, air is an electrical insulator and does not conduct electricity. However, when there is a very strong electric field the atoms and molecules in the air break apart to form negative and positive ions. The charged ions experience a force due to the electric field, and move within the electric field. This gives a **spark**.

Lightning is a very big spark. As a thundercloud charges up there is a strong electric field between the cloud and the Earth. When the field is strong enough to ionise the air, the lightning strikes.

6 Can a spark occur between two charges which are in a vacuum? Explain your answer.

7 Suggest why sparks are more likely to occur between two charged objects that are close together than between two that are far apart.

8 In a particular thunder storm, a thundercloud causes lightning to the strike the ground at a rate of once every minute. The cloud then passes over a hill. Predict what happens to the rate of lightning strikes. Justify your answer.

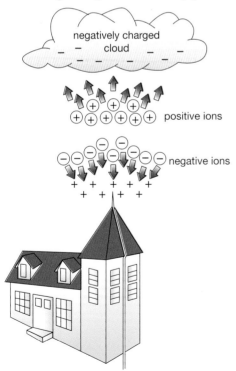

Figure 2.7 A lightning conductor is attached by a thick copper cable to a rod that is buried in the Earth below. When the air is ionised during a thunderstorm there is a path for current to travel into the ground rather than through the building itself

DID YOU KNOW?

To create an electric field that is strong enough to produce lightning, the potential difference between thunderclouds and the Earth can reach up to 35 million volts.

Electric current

Learning objectives:

- know circuit symbols
- recall that current is a rate of flow of electric charge
- recall that current (*I*) depends on resistance (*R*) and potential difference (*V*)
- explain how an electric current passes round a circuit.

KEY WORDS

coulomb
parallel
potential
   difference
resistance
voltmeter

Japanese scientists have managed to achieve an electric current of 100 000 amperes – by far the highest to be generated in the world.

Electric current and charge

Current is a rate of flow of electric charge. Electric charge is measured in **coulombs** (C). We now know that an electric current in a metal is actually a flow of electrons from negative to positive. Electric current is measured in amperes (A). This is often abbreviated to amps.

$$\text{charge flow, } Q = \text{current, } I \times \text{time, } t$$
$$\text{(in C)} \qquad \text{(in A)} \qquad \text{(in s)}$$
$$Q = It$$

Example: A wire carries a current of 1.2 A for 30 s. How much charge flows?

$Q = It = 1.2\,\text{A} \times 30\,\text{s} = 36\,\text{C}$

1 Calculate the current when 80 C of charge flows in 16 s.

2 A charge of 96 C flows in a wire carrying a current of 6 A. Calculate how long it takes for this amount of charge to flow.

Potential difference

Potential difference (pd) is the energy transferred per unit charge as charges move between two points in a circuit. It is measured in volts (V) using a **voltmeter**. A voltmeter is always connected across the component. We say the voltmeter is connected in **parallel** (Figure 2.9). Circuit symbols are shown in Figure 2.8.

For an electric current to flow, there needs to be a closed circuit (complete loop) and a source of potential difference.

Potential difference is sometimes referred to as voltage.

1 volt is the energy transferred when 1 coulomb of charge moves through a component. 1 V = 1 J/C. If a 12 V battery is used to light a lamp, each coulomb of charge going from the battery receives 12 J of energy.

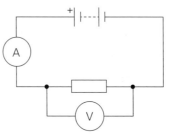

3 Write two things that are needed for there to be an electric current.

Figure 2.9 Measuring the pd across a resistor using a voltmeter

Figure 2.8 Circuit symbols

switch (open)
switch (closed)
cell
battery
diode
resistor
variable resistor
LED
lamp
fuse
voltmeter
ammeter
thermistor
LDR

4 **A 6 V battery passes a current of 1 A through a lamp for 1 minute. Calculate how much energy is transferred from the battery to the lamp.**

Current, resistance and potential difference

Electrons are 'pushed' around a circuit by a battery. They bump into the metal ions in the resistor. Energy is transferred to the metal ions, so they vibrate more and the resistor gets hotter. The increased vibrations of the ions held in their fixed lattice make it harder for the electrons to travel through the resistor, so its resistance increases (Figure 2.10).

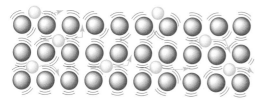

○ metal ions in the resistor

○ electrons collide with the metal ions

Figure 2.10 The movement of electrons in a wire carrying a current

The filament in a lamp connected in a circuit becomes so hot it emits light.

Note that the net direction in which the electrons move is opposite to the direction of the conventional current. The conventional current direction, from positive to negative, was well established before scientists understood that the electrons actually moved from negative to positive. So they did not change it.

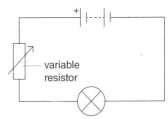

Figure 2.11 Circuit with variable resistor and lamp

5 **Explain why the bulb in Figure 2.11 gets hot.**

6 **The circuit shown in Figure 2.12 includes a variable resistor. Explain how the variable resistor can be used as a dimmer switch.**

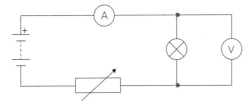

Figure 2.12

Current, *I*, **resistance**, *R*, and potential difference, *V*, are linked by the equation:

potential difference, *V* = current, *I* × resistance, *R*

(in volts, V)   (in amperes, A) (in ohms, Ω)

$$V = IR$$

Example: Calculate the potential difference across a 5 Ω resistor when the current through it is 2A.

$V = IR = 2 \times 5 = 10$ V

7 **Calculate the resistance of a car headlamp when the supply potential difference is 12 V and the current is 3 A.**

8 **Calculate the potential difference across a 6 Ω resistor when the current through it is 1.5 A.**

DID YOU KNOW?

Superconductors are materials that have an electrical resistance of zero at very low temperatures. No battery is needed to maintain a current.

ADVICE

When a current passes through a wire the wire can get hot and its resistance increases. When carrying out experiments involving current, only leave the circuit switched on for as long as you need to record measurements.

Series and parallel circuits

Learning objectives:

- recognise series and parallel circuits
- describe the changes in the current and potential difference in series and parallel circuits.

KEY WORDS

parallel circuit
series circuit

All the electrical appliances in a home are connected in a parallel circuit. If they were connected in series you would need to switch on every single appliance in order to watch TV.

Lamps in series and parallel

In the **series circuit** in Figure 2.13, the lamps are connected next to each other and form a single loop with the battery and the switch. In the **parallel circuit**, both sides of the lamps are connected to each other – a little like the rungs of a ladder.

When they are shining, the lamps connected in parallel are brighter than lamps connected in series.

If an extra lamp is added to the series circuit, then all of the lamps in the circuit are even dimmer. If an extra lamp is added to the parallel circuit, the lamps are the same brightness as before.

Series circuit

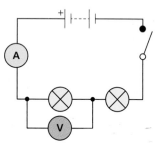

Parallel circuit

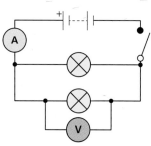

1 **None of the lamps in Figure 2.13 are currently shining. State what you would have to do to the circuits to make the lamps shine.**

2 **A third lamp is added to the series circuit. Describe what happens to the brightness of the two lamps that were already in the circuit.**

3 **Suggest what would happen if one bulb was unscrewed in each circuit.**

Figure 2.13 A series circuit and a parallel circuit

Resistors in series

In the series circuit in Figure 2.14 the current has to pass through *both* resistors, R_1 and R_2. There is nowhere else for it to go. This means that the readings on all of the ammeters are the same.

When components are connected in series, the same current flows through each component. The size of the current depends on the total resistance of the components.

For components in series, the total resistance is the sum of all of the resistances. The total resistance of the circuit in Figure 2.14 is $R_1 + R_2$.

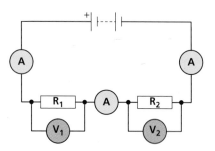

Figure 2.14 A series circuit

The potential difference of the power supply is shared between the components. This means that the reading of voltmeter V_1 plus the reading of voltmeter V_2 would equal the potential difference of the battery.

4 A motor, a lamp and a 12 V battery are connected in series. The resistance of the motor is 10 Ω and the resistance of the lamp is 20 Ω. The current through the lamp is 0.4 A.

a Determine the current through the motor.

b Calculate the total resistance of the circuit.

c If the potential difference across the motor is 4 V what is the potential difference across the lamp?

d Suggest how the current would change if the lamp was replaced with a 50 Ω resistor.

Resistors in parallel

In the parallel circuit in Figure 2.15, the current can pass through either R_1 or R_2 in the circuit. This means that the reading on ammeter A_1 is equal to the sum of the readings on ammeters A_2 and A_3.

The potential difference of the power supply is the same as the potential difference across each component. So the reading of both voltmeters V_1 and V_2 would be the same and would equal the potential difference of the power supply.

Adding a resistor in series increases the total resistance because the electric charge has to pass through another component. Adding a resistor in parallel decreases the total resistance because you are providing an alternative path for the electric charge. The total resistance of a parallel circuit is always smaller than the smallest resistance of any component.

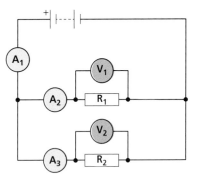

Figure 2.15 A parallel circuit

DID YOU KNOW?

A current of 1 A means that more than 6 000 000 000 000 000 000 electrons pass each point every second. This is only a small fraction of the electrons in the wire!

5 The 10 Ω motor, 20 Ω lamp and 12 V battery from Q4 are now connected in parallel. A current of 1.8 A passes through the battery and a current of 1.2 A passes through the motor.

a Determine the current through the lamp.

b Calculate the potential difference across the motor.

c Explain whether the total resistance of this circuit is greater or smaller than 10 Ω.

d A further resistor is added in parallel to the motor and the lamp. Explain what happens to the size of the current passing through the battery.

Investigating circuits

Learning objectives:

- classify materials as either conducting or insulating
- use series circuits to test components and make measurements
- carry out calculations on series circuits.

KEY WORDS

ammeter
equivalent
 resistance

When a sports match is televised it can take 24 km of electrical cabling to connect up the TV cameras and the studio. If the equipment goes wrong, engineers need to check quickly that the cables are working. How can they do this?

Circuits for testing and measuring

You can use the circuit in Figure 2.16 to check if a component or an electrical cable conducts electricity easily. The buzzer should sound when the cable or the component is attached to the terminals.

1 Explain how the circuit in Figure 2.16 works.

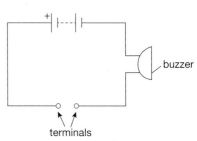

Figure 2.16 A circuit for testing components and cables

The circuit in Figure 2.17 is used to investigate the resistance of a component. You connect the component between the terminals. This circuit can be used to investigate lamps and diodes as well as components that measure temperature and light intensity.

The **ammeter** measures the current passing through the component and the voltmeter measures the potential difference across the component. Ammeters are always connected in series with the component, and voltmeters are always connected in parallel. Voltmeters have a very high resistance, which means that only a very small current flows through a voltmeter. Ammeters have a very low resistance, which means that they do not have very much effect on the resistance of the circuit.

You can change the current by altering the resistance of the variable resistor. The resistance is usually altered by moving a slider or by turning a dial.

2 An electric heater is placed between the terminals of the circuit in Figure 2.17. The ammeter reading is 2 A and the voltmeter reading is 12 V.

a Calculate the resistance of the heater.

b Describe how you could alter the circuit so that a current of 1 A passes through the heater.

REMEMBER

The resistance of a component equals the potential difference divided by the current.

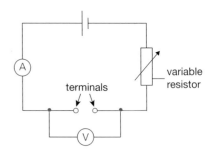

Figure 2.17 A circuit to measure the resistance of a component

Circuit calculations

These rules help you to calculate currents, potential differences and resistances in circuits.

For resistors in series, the total potential difference is the sum of the potential differences across the resistors and the current is the same through all of the resistors. The total resistance is the sum of the resistances of the resistors.

For resistors in parallel, the potential difference across each resistor is identical, but the total current is the sum of the currents that pass through each of the resistors.

3 A resistor, an ammeter and a voltmeter are connected in series with a 12 V power supply. The ammeter shows the current as 0 A. Suggest what is wrong with the circuit.

4 The potential difference across a resistor is 12 V and the current is 0.6 A. What is the value of its resistance?

5 A 5 Ω and a 7 Ω resistor are connected in series with a 6 V battery. Calculate the current in the circuit.

6 A 3 Ω and a 6 Ω resistor are connected in series with a 12 V battery. What is the potential difference across the 3 Ω resistor?

7 A 3 Ω and a 6 Ω resistor are connected in series with a battery. A current of 1 A passes through the 3 Ω resistor. Determine the potential difference of the battery.

8 A 4 Ω resistor is placed in series with another resistor and a 12 V battery. The potential difference across the 4 Ω resistor is 8 V. Determine the resistance of the other resistor.

Equivalent circuits

Sometimes you can simplify circuit calculations by replacing all the components in a circuit with a single resistor (Figure 2.18). The single resistor needs to have the **equivalent resistance** of the components in the circuit. You can then work out the current passing through the power supply.

9 A 12 V car battery is supplying a current of 60 A to the electric components in a car. The driver then switches on the sound system and a larger current passes through the battery. What happens to the equivalent resistance of the circuit?

10 A 12 Ω motor is connected in parallel with a 6 Ω lamp. Together these are connected in series with a 5 Ω resistor and a 9 V battery. The equivalent resistance of the motor, lamp and resistor is 9 Ω. Calculate the current passing through the battery.

DID YOU KNOW?

When you make a series circuit with a 1.5 V battery, a torch bulb and 2 m of connecting wire, it will take about 2 days for an electron to go around the circuit.

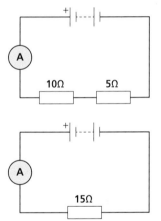

Figure 2.18 These circuits draw the same current from the battery

Circuit components

Learning objectives:

- set up a circuit to investigate resistance
- investigate the changing resistance of a filament lamp
- compare the properties of a resistor and filament lamp.

KEY WORD

filament bulb

New light bulbs have been introduced to replace the filament bulb. The so-called 'energy saver' light bulb lasts longer and is more economical to use, but it has the disadvantage that it contains mercury, which is toxic.

Measuring resistance

The circuit shown in Figure 2.19 can be used to measure the resistance of a fixed resistor. As the variable resistor is changed, the readings on the ammeter and voltmeter are recorded. A graph of current (*I*) against potential difference (*V*) is plotted. A straight line through the origin shows that current is proportional to potential difference (Figure 2.20). Such resistors are ohmic – they obey Ohm's law. Their resistance is constant. The resistance is equal to 1/gradient of an *I–V* graph. Copper wire and all other metals give this shape of graph as long as the temperature does not change.

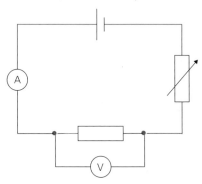

Figure 2.19 This circuit can be used to measure the resistance of a resistor

1. What does a straight line graph through the origin tell you about the quantities plotted?

2. Which meter is connected in parallel, the ammeter or the voltmeter?

The changing resistance of a filament lamp

If you switch on a light bulb using a dimmer switch, you will see that you can change the brightness of the bulb. This is because the higher the current, the higher the temperature of the filament. The hotter the bulb, the whiter and brighter the light from it becomes.

If the fixed resistor in Figure 2.19 is replaced by a **filament bulb**, the corresponding graph of current against potential difference is no longer a straight line. This is because the resistance of the filament bulb changes as its temperature changes. In Figure 2.22 the resistance is different for each value of potential difference and current.

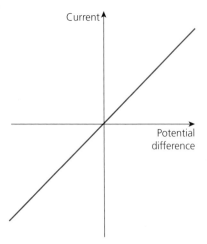

Figure 2.20 *I–V* graph for an ohmic conductor

When the filament in a bulb gets hot, two things happen. The free electrons move faster, and the metal ions in the filament vibrate more, taking up more space. As the atoms take up more space, the electrons collide with them more often, so the resistance and temperature of the bulb increase (Figure 2.21). This is why the resistance of a filament bulb increases with temperature.

The filament is made of tungsten as tungsten does not melt and evaporates very little at the typical filament bulb operating temperature of 2000 °C. An inert gas such as nitrogen or argon is usually included in the bulb to prevent the evaporation of the tungsten.

3 Explain how you know that the filament lamp is a non-ohmic conductor.

4 The filament in a bulb is made from tungsten. Tungsten, like other metals, obeys Ohm's law. Why is the *I–V* graph for a filament lamp not a straight line?

Comparing I–V graphs

A fixed resistor at constant temperature produces a straight line *I–V* graph (Figure 2.20), showing the resistance is constant. The value of the resistance is equal to the inverse of the gradient of the *I–V* graph.

The filament of a lamp is a heating element that gets so hot that it emits light. This huge temperature change means that the filament is non-ohmic – its resistance increases as the temperature increases. The *I–V* graph is a curve (Figure 2.22) – it is non-linear. The resistance of the bulb at different potential differences can only be found from instantaneous *I–V* values, not from the gradient.

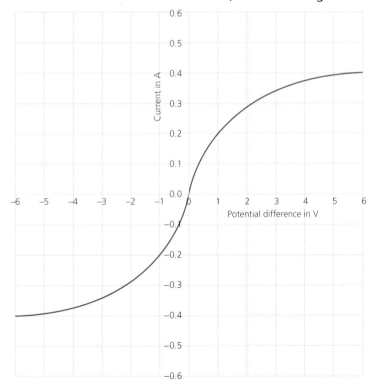

Figure 2.22 The changing slope of this graph shows that the resistance of the filament lamp increases as the current increases

5 a Calculate the resistance at 1 V and 6 V in Figure 2.22.

b Sketch a graph of resistance against time for the filament lamp immediately after it is switched on.

DID YOU KNOW?

The average light bulb has a lifetime of about 1000 hours. The tungsten used to make the filament evaporates at 2500 °C.

Figure 2.21 Coiled tungsten filament lamp

KEY INFORMATION

The resistance of a non-ohmic conductor is found from instantaneous values, not the gradient of the graph.

KEY INFORMATION

A graph of current against potential difference is called an *I–V* characteristic of the component.

REQUIRED PRACTICAL

Investigate, using circuit diagrams to construct circuits, the I–V characteristics of a filament lamp, a diode and a resistor at constant temperature

KEY WORDS

dependent
 variable
independent
 variable
ohmic conductor

Learning objectives:

*   understand how an experiment can be designed to test an idea
*   evaluate how an experimental procedure can yield more accurate data
*   interpret and explain graphs using scientific ideas.

Some components have a constant resistance. When we apply a potential difference across them current will flow. If we double the potential difference, the current doubles. We call these components ohmic. Other components do not have a constant resistance – increasing the potential difference might alter the current flow but it does not change proportionately. These are called non-ohmic. We can tell which are which by testing them, plotting graphs of the data and looking at the shape of the graph.

These pages are designed ❶ to help you think about aspects of the investigation rather than to guide you through it step by step.

Designing a circuit to test the components

The circuit shown in the figure can be used to test components to see whether their resistance changes. The component is represented by the rectangle (this is the symbol for a resistor and the component has some resistance).

The two variables are potential difference and current.

❶ Why are the ammeter and the voltmeter in the positions shown in the circuit?

❷ State the independent variable and the dependent variable.

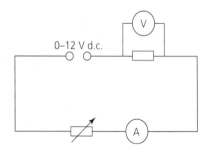

Figure 2.23 Circuit diagram for testing components

Gathering accurate data

The resistance of **ohmic conductors** can increase when their temperature increases. This is a problem with this experiment because one of the effects of passing a current through something is that its temperature increases. This has to be allowed for in the way the experiment is carried out.

3 Why should the power supply be turned off between readings?

Analysing the data

Gathering readings for potential difference and current means that they can be plotted as points on a graph. We can then draw a line of best fit for each component tested. Potential difference can be applied either way round so negative values can be gathered as well as positive. Therefore the axes need to have negative values as well as positive ones.

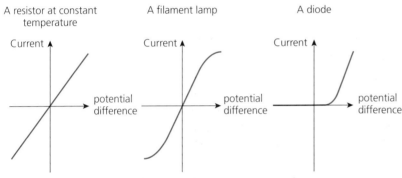

Figure 2.24

By looking at the shape of the graph we can see whether the component is ohmic. If it is, then the current will be directly proportional to the potential difference.

4 Look at the graph of the resistor. Describe the relationship between current and potential difference.

5 Look at the graph of the readings from the lamp. Explain what this shows.

6 Explain what the graph tells us about the resistance of a diode.

DID YOU KNOW?

The first diodes weren't the tiny semiconductor devices used nowadays but large glass valves that took several minutes to heat up and glowed.

REQUIRED PRACTICAL

Use circuit diagrams to set up and check appropriate circuits to investigate the factors affecting the resistance of electrical circuits, including the length of a wire at constant temperature and combinations of resistors in series and parallel

KEY WORDS

current
parallel
potential
  difference
resistance
series

Learning objectives:

- use a circuit to determine resistance
- gather valid data to use in calculations
- apply the circuit to determine the resistance of combinations of components.

We can use a circuit to determine the resistance of a component, as resistance can be calculated from measuring potential difference and current. We can then see how the resistance is affected by factors such as the length of a wire or when several components are combined.

> These pages are designed ❗ to help you think about aspects of the investigation rather than to guide you through it step by step.

Using a circuit to get useful data

The **resistance** of a component, such as a piece of wire, can be calculated using the equation:

resistance = **potential difference ÷ current**

$$R = \frac{V}{I}$$

If we set up a circuit with an ammeter and a voltmeter in it, we can record the data to calculate the resistance. The circuit we can use is shown in Figure 2.25. The rectangle represents the component being tested, such as a length of wire.

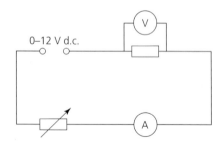

Figure 2.25 Circuit diagram for measuring potential difference across a component and the current passing through it

1. Explain what you would expect to happen to the current passing through the component as the potential difference across it is increased.

2. Describe how the potential difference across the component is altered in the circuit shown.

3. If the current passing through the component increases, describe what will happen to its temperature.

Length of a wire and its resistance

When the component being tested is a length of wire, we can find out how the resistance changes when the length is altered. For each length of wire, we can record the potential difference across the wire and current passing through it and use them to calculate the resistance. We can then plot a graph of resistance against length of the wire to show how changing the length of the wire affects the resistance.

One of the factors that can affect the accuracy of the results is temperature. As more current passes through the wire it gets hotter and this alters the resistance. It is important to avoid increasing the thermal energy store of the wire.

4 In this experiment, state what readings would be taken and what calculations would be done.

5 a Predict what effect the length of a wire will have on its resistance.

b Sketch a graph showing your prediction.

6 Suggest a good way of stopping the heating effect of the current affecting the results too much.

> **REMEMBER!**
> ....
> The greater the resistance, the higher the potential difference that is needed to make a certain current pass.

Investigating combinations of components

We can also join components together and see what the combined resistance is. We can connect them in **series**, as shown in Figure 2.26. We can then measure the potential difference across them and the current passing through and calculate the combined resistance of the resistors in series.

We can also do this with components in **parallel**, as shown in Figure 2.27. This will enable us to investigate what the combined resistance of the resistors in parallel is.

Figure 2.26 Resistors in series

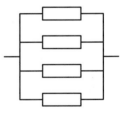

Figure 2.27 Resistors in parallel

7 a Draw circuit diagrams to show how measurements could be taken to calculate the resistance of two components in series and in parallel.

b For both series and parallel combinations, predict what you would expect the combined resistance to be, compared with the resistance of the individual components.

8 A student connected two resistors in turn to the circuit in Figure 2.25. The first resistor produced a current of 0.10 A and a pd of 6.0 V. The second resistor produced a current of 0.20 A and a pd of 3.0 V. When the student connected the resistors in series, they produced a current of 0.05 A and a pd of 3.9 V.

a Explain why the result for the resistors in series was unexpected

b Suggest a reason for the unexpected result

DID YOU KNOW?
....
The equation $V = IR$ is also known as Ohm's law. This is not strictly correct as the equation applies to any component, whereas Ohm's law only applies to certain components, and then only if the temperature is kept constant.

Control circuits

Learning objectives:

- use a thermistor and light-dependent resistor (LDR)
- investigate the properties of thermistors, LDRs and diodes.

KEY WORDS

diode
light-dependent
    resistor (LDR)
sensors
thermistor

LDRs are placed on top of street lights to turn them on when it gets dark.

Control circuits

Control circuits use components to detect changes. These components are called **sensors**.

A **thermistor** (Figure 2.28) is a temperature-dependent resistor. Its resistance changes a lot as temperature changes. At low temperatures its resistance is high. As the temperature increases its resistance decreases.

A **light-dependent resistor (LDR)** is a component whose resistance changes a lot as light intensity changes. When it is light the resistance of the LDR is low. When it is dark the resistance of the LDR is very high.

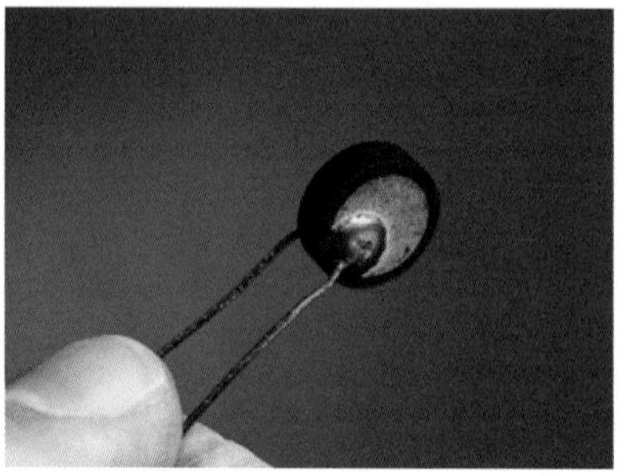

Figure 2.28 Thermistor

1️⃣ **What property of an LDR changes as the light level changes?**

2️⃣ **What happens when the temperature of a thermistor increases?**

The properties of thermistors

The resistance of a thermistor decreases as the temperature increases. Each thermistor has its own characteristics, but the resistance of a typical thermistor changes from 2000 Ω at −20 °C to 200 Ω at 20 °C.

A thermistor is made from a semi-conductor. A semi-conductor is neither a good conductor nor an insulator. When a semi-conductor is heated it can conduct more easily. This is because the rise in temperature releases more free electrons to carry the current. The higher the temperature the lower the resistance (Figure 2.29).

The resistance of a thermistor is highest when cold. It can be used to:

- turn on a heater when it gets cold, either in the house or in a greenhouse
- act as a fire alarm
- keep a fish tank from becoming too cold.

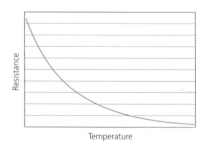

Figure 2.29 Graph of resistance against temperature for a thermistor

3 Explain how you know that a thermistor is not an ohmic conductor.

4 Explain how Figure 2.29 shows that a thermistor's resistance decreases when the temperature increases

The properties of light-dependent resistors

In bright sunlight an LDR has a resistance of about 100 Ω. When it is dark the resistance of the LDR becomes very large (Figure 2.30). It can be over 10 MΩ (10 000 000 Ω) in the dark.

An LDR connected in series with a battery and an ammeter can be used to make a simple light meter.

When it is bright the resistance of the LDR is low and the reading on the ammeter is high. When it is dark the resistance of the LDR is high and the reading on the ammeter is low.

This can be used by a cricket umpire to decide whether it is too dark to carry on playing safely.

5 Describe what happens to the current if you cover and uncover an LDR when a bright light is shining on it.

The properties of a diode

A **diode** is a component that only allows a current to flow in one direction – the direction in which the arrow points in the circuit symbol (Figure 2.31).

The current–potential difference characteristics for a diode (Figure 2.32) can be found using a circuit similar to the one used to draw the I–V graph for a filament lamp.

Most diodes start to conduct when the potential difference across them is about 0.6 V. The steep slope of the graph shows that current passes easily through the diode when it is conducting. With a negative potential difference, very little current flows.

6 Explain what the I–V graph for a diode tells you about the resistance of the diode when the potential difference is negative.

7 Explain why you often need another resistor to protect a diode in a circuit.

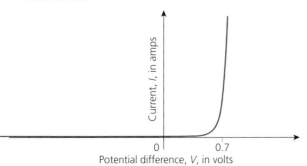

Figure 2.32 I–V graph for a diode

DID YOU KNOW?

Some incubators for newborn babies use thermistors that detect a 0.1 °C change in temperature.

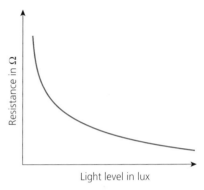

Figure 2.30 Change in resistance with light level in an LDR

REMEMBER!

The resistance of an LDR is Low in the Light.

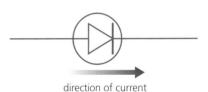

direction of current

Figure 2.31 The arrow on the diode symbol tells you which way the current passes through it

Electricity in the home

Learning objectives:

- recall that the domestic supply in the UK is a.c. at 50 Hz and about 230 V
- describe the main features of live, neutral and earth wires.

A circuit-breaker (Figure 2.33) is a resettable fuse. It has replaced the wire fuse in the main fuse box as it can be reset at the flick of a switch.

Domestic electricity supply

A cell or battery has two terminals, positive and negative. The current is d.c., which means that it always passes in the same direction.

Mains electricity differs in two ways (Figure 2.34):

- The current alternates, that is it changes direction. It has a frequency of 50 Hz.
- It has a much higher potential difference (about 230 V).

Figure 2.33 Circuit-breaker

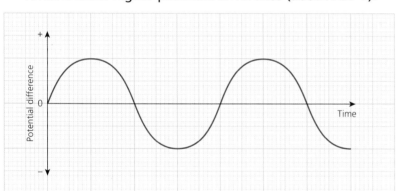

Figure 2.34 The potential difference variation from the mains

1. Describe the difference between a direct and an alternating potential difference.

2. The mains electricity in the USA has a potential difference of 120 V and a frequency of 60 Hz. Describe how Figure 2.34 would change for the USA.

Connecting a three-pin plug

Live, neutral and **earth** wires can be seen in a plug (Figure 2.35). If the appliance is working properly, there should be no current in the earth wire.

The three wires inside the mains cable are colour-coded:

- brown is connected to the live terminal (L)
- blue is connected to the neutral terminal (N)
- green/yellow stripe is connected to the earth terminal (E).

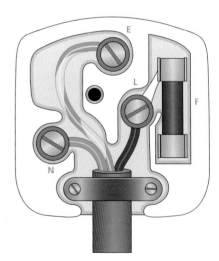

Figure 2.35 What colour is the earth wire?

3 Explain why most electric wires in the home are covered in plastic.

4 Suggest why the three wires have to be very different colours.

Live, neutral and earth wires

Our bodies are at Earth potential – there is no potential difference between our bodies and the Earth. If the casing of a faulty appliance becomes live, the potential difference between it and Earth is 230 V. If a person touches it, there is a complete circuit and current passes through the person's body to Earth. The person receives an electric shock. If the appliance is earthed it is connected to the Earth or ground (Figure 2.36). Any charge in it flows safely down to the ground through the low-resistance earth wire. This stops a person receiving an electric shock if they touch the live case of a faulty appliance.

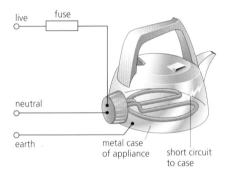

Figure 2.36 The arrangement of wires in a metal-cased appliance

Figure 2.37 Double insulation symbol

The mains electricity supply usually comes from a power station. Two wires connect a house to a power station – live and neutral.

- The live wire carries a high potential difference into and around the house. The **fuse** in the plug (F in Figure 2.35) is always connected to the live wire.
- The neutral wire provides a return path to the local sub-station. The neutral wire is earthed and so is at, or close to, earth potential (0 V). There is no current in the neutral wire until an electrical appliance is connected.
- The earth wire is a safety wire. It is connected to the metal case of an appliance to prevent it becoming charged if touched by a live wire. It provides a low-resistance path to the ground. There is normally no current in it.

5 Explain why a battery-powered torch has two connections to its power supply but a mains lamp has three.

6 Describe the function of the earth wire. Explain whether the current is high or low when there is a fault and how this makes the appliance safe.

DID YOU KNOW?

A double-insulated appliance does not need an earth connection (Figure 2.37). It has a plastic case with no electrical connections to it, so the case cannot become live.

REMEMBER!

A fuse is always connected in the live wire. If the current becomes higher than the fuse's rating the wire in the fuse will melt and switch off the circuit.

Transmitting electricity

Learning objectives:

- describe how electricity is transmitted using the National Grid
- explain why electrical power is transmitted at high potential differences
- understand the role of transformers.

KEY WORDS

National Grid
transformer

In the USA, the potential difference used is 120 V instead of 230 V as in the UK. This means the wires have to carry about twice the current to make appliances work.

The National Grid

When electricity was first delivered to homes, each town had its own local power station. If the power station broke down then the town did not receive any electric power. Each power station produced electricity at a different potential difference. This meant that electric appliances, such as light bulbs, would only work in one town. If you moved house you would have to buy new ones.

The **National Grid** is a collection of power cables and **transformers** that connect power stations to factories and houses across Great Britain. Everything is connected up in a grid so that electrical power can be transferred along many different routes. This means that the electrical power in your house can come from lots of different power stations.

Figure 2.38 Power cables connected in the National Grid

1 State an advantage of connecting your house to the National Grid rather than just connecting it up to a local power station.

2 Explain what the words National and Grid mean when referring to the transmission of electrical power.

Changing the potential difference

The potential difference of the domestic UK mains supply is 230 V. This means that there is a potential difference of 230 V between the live wires and the earth wires in a house.

Electrical power is transmitted across the country at 400 000 V. The potential difference between a live power line and the earth is very large.

3 Explain why it would not be a good idea for the potential difference to be 400 000V in a domestic electricity supply.

An electric current passing through the National Grid's power cables heats up the cables. Energy is being transferred to the thermal energy stores of the surroundings rather than useful energy stores in houses and factories.

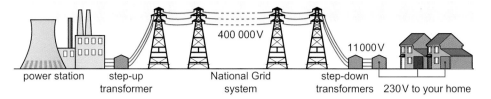

400 000 V

11 000 V

| power station | step-up transformer | National Grid system | step-down transformers | 230 V to your home |

Figure 2.39 The National Grid

The higher the current, the more energy is wasted and the National Grid becomes less efficient. For the same electrical power, increasing the potential difference reduces the current. So using a high potential difference makes the National Grid more efficient.

4 Suggest why it is important that energy wasted by the National Grid is minimised.

5 How does increasing the potential difference make the National Grid more efficient?

Transformers

Transformers are devices that can change the potential difference. They are made from two coils of wire with some iron passing between them. Varying the number of turns on each coil changes the potential difference.

Step-up transformers increase the potential difference. They are connected between the power stations and the National Grid (Figure 2.39). Step-down transformers decrease the potential difference.

Figure 2.40 Step-up transformers are installed close to the power stations

6 State where step-down transformers are connected in the National Grid.

7 Transformers have no moving parts. Suggest why this means that they are very efficient.

8 Suggest why step-up transformers are placed as close to power stations as possible (Figure 2.40).

9 Explain why the birds on the high potential difference power lines in Figure 2.41 are safe.

Figure 2.41 Birds on power lines

Power and energy transfers

KEY WORD
....
power

Learning objectives:

- describe the energy transfers in different domestic appliances
- describe power as a rate of energy transfer
- calculate the energy transferred.

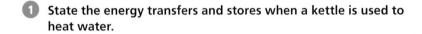

Much more energy is transferred when you heat a room than when you light a room.

Energy transfers

Electrical devices that we use every day are designed to transfer energy. All electrical appliances transfer energy. But different appliances transfer energy in different ways. They transfer energy from the a.c. mains supply or stores such as batteries to stores such as the kinetic energy store of a motor or the thermal energy store of hot water in a kettle.

1 **State the energy transfers and stores when a kettle is used to heat water.**

Figure 2.42 A hairdryer

Power and energy transferred

Consider a hairdryer (Figure 2.42). The hairdryer receives 1500 J of energy each second from the mains supply. It transfers 1500 J of energy per second from the mains supply to two main energy stores: thermal energy stored in the air and kinetic energy in the moving air.

The **power** is the amount of energy transferred each second. The units are joules per second, or watts (W). As the hairdryer transfers 1500 J each second, its power is 1500 W or 1.5 kW.

2 **An electric drill (Figure 2.43) connected to the a.c. mains transfers 400 J every second.**

   a **Describe how the drill transfers energy and the stores it transfers energy to.**
   b **State the power of the drill.**

Figure 2.43 An electric drill

The amount of energy transferred by an appliance depends on its power and the length of time it is used. We can use this equation to calculate the amount of energy transferred.

energy transferred (in joules, J) = power (in watts, W) × time (in seconds, s)

$E = Pt$

Example:

The hairdryer in Figure 2.42 is used for 5 minutes. Calculate the total energy transferred by the hairdryer.

$E = Pt$

  = 1500 W $\times$ (5 $\times$ 60) s

  = 450 000 J (or 450 kJ)

3 An electric oven with a power rating of 2.5 kW is switched on for 45 minutes.

Calculate the total energy transferred by the oven.

4 A hairdryer transfers 10 000 J of energy from the a.c. mains supply in 5 s. Calculate its power.

Charge and energy transferred

When charge flows in a circuit, electrical work is done. We can calculate the amount of energy transferred by electrical work using the equation:

energy transferred, E = charge flow, Q $\times$ potential difference, V

(in joules, J)　　　　　(in coulombs, C)　　　　(in volts, V)

$E = QV$

Example:

A charge of 50 C flows through a device with a potential difference across the device of 12 V.

Calculate the total energy transferred.

$E = QV$

  = 50 C $\times$ 12 V

  = 600 J

5 A charge of 30 C flows through a TV which is connected to the mains supply of 230 V.

Calculate the total energy transferred.

6 A device transfers a total of 1800 J with a charge of 75 C.

Calculate the potential difference across the device.

7 A home appliance has a power of 1150 W and is switched on for 5 minutes.

a Determine how much charge flows through the appliance.
b Calculate the amount of charge that passes through a 415 V factory appliance when the same amount of energy is transferred.

DID YOU KNOW?

Homes and offices are supplied with electricity at 230 V, but small factories have an 11 000 V supply and large factories 33 000 V.

Calculating power

KEY WORD
....
power

Learning objectives:

- calculate power
- use power equations to solve problems
- consider power ratings and changes in stored energy.

We use the word power a lot – 'a powerful idea' or 'a powerful piece of music'. However, in science, we use it in a very specific way – as a measure of how quickly energy is transferred. Unlike ideas or music, this power can be calculated.

Calculating power

The **power** transfer in any component is related to the potential difference across it and the current passing through it. We can calculate power using the equation:

power, P = potential difference, V × current, I

(in watts, W)   (in volts, V)   (in amps, A)

$P = VI$

When a current passes through a resistor, such as a kettle element, it has a heating effect. Work is done by the electrons which is transferred to thermal energy in the element.

As $V = IR$, so $P = (IR) \times I = I^2R$

So the power transfer is also given by the equation:

power   =   (current)²   ×   resistance

(in watts, W)   (in amps, A)²   (in ohms, Ω)

$P = I^2R$

1 **An electric heater takes a current of 4 A when connected to a 230 V supply.**

   a Calculate its power.

   b Calculate its resistance.

2 **A lamp has a power of 36 W when connected to a 12 V supply.**

   a Calculate the current through the lamp.

   b Calculate the resistance of the lamp.

Heating up

We can use equations for power and energy transfer to solve problems.

Example: Jo boils a kettle of water to make a cup of tea (Figure 2.44). The kettle has a power rating of 2.4 kW.

The mains supply is 230 V. Calculate:

a the current in the kettle element

b the resistance of the kettle element.

Answer:

a $P = VI$

   Rearrange the equation to make I the subject:

   $I = P \div V$
   $= 2400\ \text{W} \div 230\ \text{V}$
   $= 10.4\ \text{A}$

b $P = I^2 R$ resistance of element, $R = P \div I^2 =$
   $2400 \div (10.4)^2 = 22.2\ \Omega$

Figure 2.44 An electric kettle

Example:

An electric kettle connected to the 230 V mains supply draws a current of 10 A. It contains 2 kg of water. Calculate the rise in temperature of the water when the kettle is switched on for 1 minute.

The specific heat capacity of water is 4200 J/kg °C

Power of kettle, $P = VI = 230\ V \times 10\ A = 2300\ W$

Energy transferred in 1 minute (60 s) $= Pt$

$= 2300\ W \times 60\ s = 138\ 000\ J$

If we assume all this energy is given to the 2 kg of water in the kettle we can calculate the rise in temperature, $\Delta\theta$. Also, see topic 1.5.

Energy transferred to water $= mc\Delta\theta$

$138\ 000\ J = 2\ kg \times 4200\ J/kg\ °C \times \Delta\theta$

$\Delta\theta = 138\ 000 \div (2 \times 4200) = 16.4\ °C$

3 In the example above it was assumed that all the energy transferred by the electric current is transferred into thermal energy of the water. Suggest what other stores the energy might be transferred to.

4 Al has an outdoor swimming pool. It is 15 m long, 10 m wide and 2 m deep. The density of water is 1000 kg/m³.

a Calculate the mass of water in the pool.

b Al wants to warm the water from 17 °C to 22 °C. Calculate how much energy is transferred.

c The power of the heater is 2 kW. Calculate how long it takes to raise the temperature of the water.

Figure 2.45 Energy is transferred between stores as this car moves

Changes in stored energy

A more powerful appliance can transfer energy more quickly.

A **battery-operated toy car** with a power rating of 5 W transfers 5 J every second. Energy is transferred from the store of chemical energy in the battery to the store of kinetic energy in the toy car. Although the total amount of energy is conserved, the decrease in the energy store of the battery does not equal the increase in the kinetic energy store of the car. The toy does work against friction, so the energy that is dissipated is transferred to the thermal energy store of the surroundings.

An **electric cooker** with a power rating of 2 kW transfers 2 kJ every second. The energy transferred by the electric current increases the thermal energy stored in the food, the saucepan and the surroundings.

Figure 2.46 Cooking involves energy being transferred from one store to another

5 Describe how different stores of energy change when these appliances are in use:

a A microwave oven with a power rating of 800 W.

b A vacuum cleaner with a power rating of 1.6 kW.

What's the difference between potential difference and current?

Learning objectives:

- understand and be able to apply the concepts of current and potential difference
- use these concepts to explain various situations

KEY WORDS

charge
current
energy transfer
potential
  difference
power
resistance

When we're exploring and using electrical circuits it's useful to be able to measure things. This enables us to explain why electricity is sometimes really safe and sometimes lethally dangerous. It isn't simply a case of saying that 'there's more electricity in one than the other'. You may well have seen a Van de Graaff generator being used. This produces sparks that will jump across several centimetres of air, yet science teachers will sometimes demonstrate how they can make people's hair stand on end.

Potential difference and current

To understand what is going on, we need to think about two quantities, **potential difference** and **current**. These are not the same; understanding the difference will help make sense of a lot of things to do with circuits.

Current is a flow of **charge**. If you rub a balloon on a woollen jumper it becomes charged. The charge stays in the balloon (which is why it's called static – it doesn't move). As soon as it moves – we have a current. If you set up a simple circuit with battery, wires and a bulb, the bulb lights because there is a current in the circuit. Current is a flow of electrons moving through the wires. Current is measured in amperes (A).

Potential difference is the work done in moving a unit of charge. It is an indication of how much energy is transferred to a unit charge when charge moves between two points, such as between the terminals of a battery. Potential difference is measured in volts (V). Potential difference is often referred to as voltage and both terms can be used, but the appropriate scientific term is potential difference. A 1.5 V battery does not transfer much energy to each unit of charge. In Figure 2.47, if you touch the two terminals you won't feel anything. A 12 V battery (such as that used in a car) transfers more energy to each unit of charge and the mains electricity supply (230 V) transfers much more energy to each unit charge.

Figure 2.47 A 1.5 V battery

1. If you pull a nylon jumper over your head in a darkened room you can sometimes feel something crackling and see sparks. What is happening?

2. Trucks have electrical systems that run on 24 V. How could you produce this using two car batteries?

The difference between potential difference and current

Potential difference is the **energy transferred** per coulomb of charge between two points. Current is the flow of electrical charge. The size of the electric current is the rate of flow of electrical charge.

The bulb in the torch shown in Figure 2.48 is designed to use 4.5 V and carry 750 mA of current. The energy transferred by the current is enough to make the bulb glow, but not so high that the bulb would blow (break).

Figure 2.48 A torch

The electric heater shown in Figure 2.49 is designed to use mains potential difference (230 V) and carry 8 A. Each coulomb of charge is carrying more energy and there is also more electrical charge flowing per second.

When lightning strikes, both the potential difference and current are huge. A thunderstorm can generate a potential difference of up to 500 000 V and the current from lightning can reach thousands of amps.

Figure 2.49 An electric heater

So why can someone touch the charged dome of a Van de Graaff generator and not be in danger? The potential difference is high (even a small generator designed for schools can produce up to 100 000 V) but the discharge current is tiny, usually only a few milliamps. The high potential difference produces the spectacular effects but the low current means it's safe.

Why are potential difference and current important ideas?

Figure 2.50 An electric oven

Electrical **power** shows how quickly energy is being transferred; it is calculated from potential difference times current: $P = VI$. Increasing either V or I means more power.

Resistance shows how hard the current is being opposed. A greater resistance means that there is less current (if the potential difference is the same). Sometimes we want a high current to pass through a component, such as the heating element in an electric oven (Figure 2.50). An electric oven can have a power rating of about 4 kW and operate at up to 30 A. On other occasions we want the current to be very small, such as for a light bulb. The electric oven and light bulb both operate from the mains, but the light bulb has a much higher resistance. Resistance is calculated from potential difference divided by current: $R = \frac{V}{I}$.

3 Is it true to say that the highest potential differences are always the most dangerous?

4 Why do electricians sometimes say that 'It's the volts that jolts but the current that kills'?

5 What is the power rating of a bulb working on the potential difference of 230 V and drawing a current of 0.05 A?

6 What would the resistance of this bulb be?

MATHS SKILLS

Using formulae and understanding graphs

Learning objectives:

- recognise how algebraic equations define the relationships between variables
- solve simple algebraic equations by substituting numerical values
- describe relationships expressed in graphical form.

Mathematics helps you to express abstract concepts and ideas in a more concrete and helpful way.

Algebraic equations and relationships

Algebraic equations show relationships between variables, which are represented as letters. For example, the relationship between potential difference (V), current (I) and resistance (R):

$V = IR$, where:

V = potential difference in volts, V

I = current in amps, A

R = resistance in ohms, Ω

From the equation, you can see that if the resistance R is constant, the bigger the potential difference V between two points in a circuit, the bigger the current I. The current through a conductor between two points is **directly proportional** to the potential difference across the two points. This is written mathematically as:

$V \propto I$, where the symbol $\propto$ means 'proportional to'.

1 **a** Calculate the potential difference across a component of resistance 100 Ω when the current is 10 A.

  b Calculate the potential difference across a component of resistance 3 kΩ when the current is 5 A.

2 Show how you would rearrange the equation $V = IR$ to calculate the resistance R of a component.

Power and energy transfer in circuits

The equation for power is:

power = potential difference × current

> **REMEMBER!**
>
> To express very large quantities, use larger multiples or smaller submultiples of the base unit. For example, 1000 Ω is usually written as 1 kΩ, and 2500 Ω can be written as 2.5 kΩ. Here, 'k' means kilo, which is 10^3 or 1000. But remember – in calculations, always use the base unit (in this example, ohms): 1000 Ω and 2500 Ω.

and the algebraic equation is:

$P = VI$, where:

P = power in watts, W

V = potential difference in volts, V

I = current in amps, A

The same relationship can also be expressed by substituting IR for V, as $V = IR$. The expression becomes:

$P = (IR) \times I$

which can be written as $P = I^2R$.

3 Calculate the power of a motor supplied with a potential difference of 230 V and a current of 5 A.

4 How would you rearrange the equation to calculate the current I when you are given the potential difference V and the power P?

5 Calculate the power of a motor supplied with a current of 5 A if the resistance is 1000 Ω.

Describing relationships expressed in graphical form

Figure 2.51 shows that there is a **linear relationship** between potential difference and current through an ohmic conductor at constant temperature – the graph is a straight line. Because the straight line goes through the origin, we can also say that current is directly proportional to the potential difference, $V \propto I$.

We can insert a **constant of proportionality** (k) to describe the relationship between V and I:

$V = kI$

In Figure 2.51, the reciprocal of the slope or gradient of the line (1/gradient) is equal to the constant of proportionality, which in this graph is the resistance in the circuit, R. The straight line indicates that the resistance is constant (it does not change).

The relationship between potential difference and current flowing through a filament lamp is different (Figure 2.52). Initially, at low values of V and I, the gradient is straight, but then it quickly becomes a curve. This indicates that the gradient is changing, which means that the resistance is changing.

The resistance of a filament lamp increases as the temperature of its filament increases. As a result, the current flowing through a filament lamp is not directly proportional to the potential difference across it.

6 What does the slope of a straight-line graph indicate?

7 Looking at a graph, how would you know if the relationship is not directly proportional?

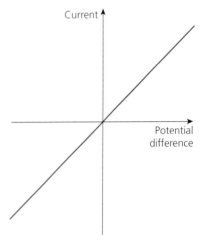

Figure 2.51 *I–V* graph for an ohmic conductor at constant temperature

KEY INFORMATION

Note also that the values of potential difference and current can also be negative, so the line stretches back into the negative quadrant.

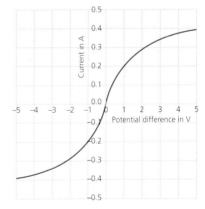

Figure 2.52 *I–V* graph for a filament lamp

Check your progress

You should be able to:

Describe how insulating materials can become charged → Recall that there are two types of charge and that like charges repel and unlike charges attract → Explain how a person can get an electric shock and explain static electricity in terms of electric fields

Recall that an electric current is a flow of electrical charge and is measured in amperes (A) → Remember that charge is measured in coulombs (C) and recall and use the equation $Q = It$ → Explain the concept that current is the rate of flow of charge. Rearrange and apply the equation $Q = It$

Recognise and use electric circuit symbols in circuit diagrams → Draw and recognise series and parallel circuits. Compare the brightness of lamps connected in series and parallel → Recall that the current in a series circuit is always the same and that the total current in a parallel circuit is the sum of the currents through each branch

Recall that the current through a component depends on the resistance of the component and the potential difference across it → Recall and apply the equation $V = IR$ and for series circuit $R_{total} = R_1 + R_2$ → Explain the effect of adding more resistors to series and parallel circuits

Set up a circuit to investigate the relationship between V, I and R for a fixed resistor → Draw I–V graphs for a fixed resistor → Analyse and interpret I–V graphs for a fixed resistor

State the main properties of a diode, thermistor and light-dependent resistor (LDR) → Describe the behaviour of a thermistor and LDR in terms of changes to their resistance → Describe applications of diodes, thermistors and LDRs and explain their uses

Draw I–V graphs for filament lamps → Explain the properties of components using I–V graphs → Use I–V graphs to determine if the characteristics of components are ohmic or non-ohmic

Recall that cells and batteries produce low-voltage direct current → Recall that domestic supply in the UK is 230 V a.c. and 50 Hz → Explain the difference between direct and alternating potential difference

Identify live, neutral and earth wires by their colour-coded insulation → Explain why a live wire may be dangerous even when a switch in the main circuit is open → Explain the dangers of providing any connection between the live wire and earth or our bodies

Recall that the National Grid is a system of cables and transformers linking power stations to consumers → Describe how step-up and step-down transformers change the potential difference in the National Grid → Explain why electrical power is transmitted at high voltages in the National Grid

Understand that everyday electrical appliances bring about energy transfer → Recall and use the equation energy transferred $E = Pt$ → Recall and apply the equation energy transferred $E = QV$

Recall that power is measured in watts (W) and 1 kW = 1000 W → Recall and use the equation $P = V \times I$ → Recall and apply the equation $P = I^2R$

Worked example

The diagram below shows a circuit used to investigate the resistance of a piece of thin wire.

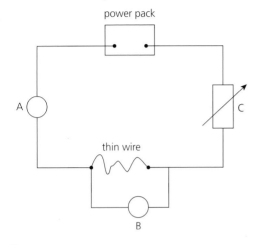

1 **Name the three components in the circuit labelled A, B and C.**

A is an ammeter, B is a voltmeter and C is a thermistor.

> A and B are correct, but C is a variable resistor.

2 **Give the purpose of component C.**

To regulate the temperature.

> Component C is used to change the potential difference across the test material.

3 **The student recorded the potential difference across the thin wire and the current passing through it. She plotted her results on a graph. Explain which variable she should plot on each axis.**

She should plot potential difference on the x-axis and current on the y-axis.

> This is correct, but the explanation has not been given. Potential difference is plotted on the x-axis because it is the independent variable, and current is plotted on the y-axis because it is the dependent variable.

4 **The student increased the potential difference to 12 V. Explain what you think would have happened.**

The wire gets very hot.

> The wire got hot because of the large current passing through it. Large currents transfer lots of energy.

5 **Explain how you would expect the graph to look if the wire had been replaced by a filament lamp.**

The same

> The line would initially be straight but then curve with a shallower gradient as the temperature of the filament increases.

End of chapter questions

Getting started

1. Write the colour of the insulation on the earth wire in a three-pin plug. `1 Mark`

2. Draw the circuit symbol for a voltmeter. `1 Mark`

3. State whether the ammeter should be connected in series or parallel when measuring an electric current. `1 Mark`

4. Describe what we mean by the **National Grid**. `2 Marks`

5. Calculate the current when 100 C of charge flows in 20 s. `2 Marks`

6. Describe how you could charge a balloon so that it will stick to the wall. `1 Mark`

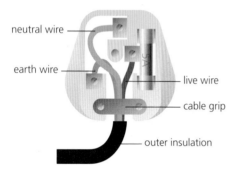

7. Describe what is wrong with the wiring in the plug in the diagram above. `2 Marks`

Going further

8. Draw the circuit symbol for a thermistor. `1 Mark`

9. Explain the effect of putting a thermistor in a circuit. `2 Marks`

10. A charged object produces an electric field. Describe what is meant by an electric field. `2 Marks`

11. Two balloons are hung down next to each other by cotton threads. Describe what will happen if both balloons are positively charged. `1 Mark`

12. Suggest what will happen to the brightness of the lamps and the current in a series circuit if an additional lamp is added. `2 Marks`

13. A 2 kW kettle is attached to a 230 V mains supply. Calculate the current through the kettle. `2 Marks`

More challenging

14. What happens to the resistance of an LDR when the light level is decreased? `1 Mark`

15. Explain how a person gets an electric shock if they are charged and then earthed. `1 Mark`

16. Explain how a fuse acts as a safety device. `2 Marks`

17. Draw a circuit diagram of a circuit you could use to find the resistance of a short piece of nichrome wire. `2 Marks`

18 Calculate the current through a lamp of resistance 8 Ω when connected to a 12 V battery. `2 Marks`

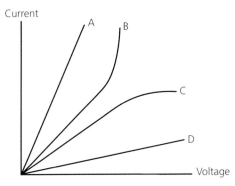

19 The current–voltage characteristics for four components, A–D, were investigated and the results drawn on the graph above. Which component was a filament lamp and which was a high-value resistor? `2 Marks`

Most demanding

20 Give power as an equation in terms of current and resistance. `1 Mark`

21 Explain the effect of adding resistors in series and in parallel on the total resistance of a circuit. `2 Marks`

22 Calculate the energy transferred in 1 minute by a 60 W light bulb connected to a 230 V mains supply. `2 Marks`

23 Four appliances were switched on for various times. Incomplete information about each and how long they were on for is shown in the table below. Which appliance transferred the most energy in the time given and how much energy was transferred? `2 Marks`

Appliance	Power rating (W)	Current (A)	Voltage (V)	Time left on (s)
Kettle	3000	12		20
Microwave	920		230	60
Torch		5	12	200
Food mixer		6	230	50

24 In certain places, the National Grid transmits electricity at a potential difference of 44 kV. Evaluate the use of a high potential difference when transmitting electrical power. `3 Marks`

`Total: 40 Marks`

PARTICLE MODEL OF MATTER

IDEAS YOU HAVE MET BEFORE:

ENERGY CHANGES AND TRANSFERS

- The concept of heating and thermal equilibrium
- The temperature difference between two objects leads to energy transfer from the hotter to the cooler one

CHANGES OF STATE

- There are similarities and differences between solids, liquids and gases
- Energy has to be supplied to melt a solid or to boil a liquid

PRESSURE OF A GAS

- Atmospheric pressure decreases as the height above sea level increases.
- In a gas the particles are far apart and move freely.
- Pressure is measured in pascals (Pa). 1 Pa = 1 N/m^2.

PARTICLE MODEL

- Every material is made of tiny moving particles.
- The higher the temperature the more kinetic energy the particles have and the faster they move.

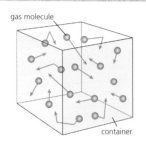

gas molecule

container

IN THIS CHAPTER YOU WILL FIND OUT ABOUT:

WHAT USES ARE MADE OF THE HIGH SPECIFIC HEAT CAPACITY OF WATER?

- The specific heat capacity, c, is the energy required to raise the temperature of 1 kg of an object by 1 °C. The unit for c is J/kg °C.
- Water has a very high specific heat capacity. A lot of energy is needed to heat the water in a radiator or hot water bottle.
- The internal energy of the system is the total amount of kinetic energy and potential energy.

WHAT ARE THE SPECIFIC LATENT HEAT OF VAPORISATION AND THE SPECIFIC LATENT HEAT OF FUSION?

- A particle model is used to show that mass is conserved when changing state.
- The specific latent heat of fusion is the energy required to change 1 kg of an object from a solid to a liquid without a change in temperature.
- The specific latent heat of vaporisation is the energy required to change 1 kg of an object from a liquid to a gas without a change in temperature.

WHAT HAPPENS TO THE PRESSURE OF A GAS WHEN IT IS HEATED, KEEPING THE VOLUME CONSTANT?

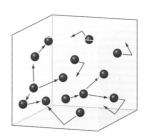

- An increase in temperature increases the kinetic energy of the gas particles.
- The particles move faster, colliding more often and with greater force on the walls of their container. The pressure of the gas increases.

HOW ARE PRESSURE AND VOLUME CONNECTED?

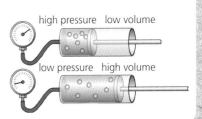

high pressure   low volume

low pressure   high volume

- When the mass and temperature are constant, the pressure multiplied by the volume is a constant.
- Increasing the volume of a gas can decrease the pressure.

Density

Learning objectives:

- use the particle model to explain the different states of matter and differences in density
- calculate density.

Inspectors measure the density of milk and beer to see whether they have been watered down.

States of matter

A **solid** has a fixed size and shape. The particles vibrate to and fro but cannot change their positions. They are held together by strong forces of attraction called **bonds**.

A **liquid** has a fixed size but not a fixed shape. It takes the shape of its container. The bonds between particles are less strong than in a solid. The particles are close together and attract each other. They move around but have no regular pattern.

A **gas** has no fixed size or shape. Its particles move randomly and fill the space available. The particles are well spaced out and virtually free of any attractions. They move quickly, colliding with each other and the walls of their container.

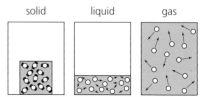

Figure 3.1 **Particle model** of particles in a solid, liquid and gas

1. Which state does each of the following describe?

   a   Fixed shape and volume.

   b   Particles move around freely at high speed.

   c   Fixed volume but no fixed shape.

Differences in density

Some people say 'lead is heavier than iron'. What they mean is that a piece of lead is heavier than a piece of iron of the same volume. **Density** compares the mass of materials with the same volume. Lead is denser than iron.

Densities of different materials

Substance	Density in kg/m³
gold	19 000
iron	8000
lead	11 000
cork	250
mercury	13 600
water	1000
petrol	800
air	1.3

In general, the liquids in the table on page 86 have densities lower than solids but higher than gases, but there are exceptions. The reason for the lower density is that the particles in a liquid are not as tightly packed as the particles of a solid (Figure 3.1).

The particles in a gas are more spread out, unless compressed. Gases have very low densities.

Mercury and cork do not follow the pattern. Mercury is a very dense liquid. Cork is a solid with a low density.

2 Explain why solids usually have a higher density than liquids and gases.

3 Explain why gases have a low density.

Density

Density is the mass of unit volume of a substance. It is given by the equation:

$$\text{density} = \frac{\text{mass (in kg)}}{\text{volume (in m}^3)}$$

$$\rho = \frac{m}{V}$$

where mass is in kg, volume in m^3 and density is in kg/m^3.

Density is also given in g/cm^3 where mass is in g and volume is in cm^3.

$1\ g/cm^3 = 1000\ kg/m^3$

Example: Calculate the mass of 3 m^3 of water. The density of water is 1000 kg/m^3.

$$\rho = \frac{m}{V}$$

Rearrange the equation to give:

$m = \rho V$

$\quad = 1000\ kg/m^3 \times 3\ m^3$

$\quad = 3000\ kg$

4 a Calculate the density of a 5400 kg block of aluminium with a volume of 2 m^3.

　b Calculate the mass of steel having the same volume as the aluminium block. Density of steel = 7700 kg/m^3.

　c Explain why aluminium is used to build aeroplanes rather than steel (Figure 3.2).

5 Calculate the mass of air in a room 5 m by 4 m by 3 m. The density of air is 1.3 kg/m^3.

6 Suggest why cork floats in water, but iron sinks.

7 Explain what happens to the density of air in a bicycle tyre when someone sits on the bicycle.

8 Explain why a density of 1 g/cm^3 is the same as a density of 1000 kg/m^3.

Figure 3.2 Materials used in aeroplanes have to be light but strong.

REQUIRED PRACTICAL

To investigate the densities of regular and irregular solid objects and liquids

Learning objectives:

- interpret observations and data
- use spatial models to solve problems
- plan experiments and devise procedures
- use an appropriate number of significant figures in measurements and calculations.

There is a great story about how scientists learnt how to measure density. **Density** is worked out knowing the mass of an object and the volume it occupies. Measuring the volume is easy enough if the object is a regular shape such as a cube but what if it was, say, a crown?

This problem was given to Archimedes, a clever man who lived thousands of years ago in Greece. The king had ordered a new crown to be made but suspected the craftsman had mixed a cheaper metal with the gold. Measuring the density would reveal if the gold was pure, but how could the volume be found? Archimedes realised that by carefully immersing the crown in a full can of water, the water that overflowed would have the same volume as the crown.

Apparently he shouted 'Eureka!' ('I have found it!') and the cans used in practical investigations are known as Eureka cans in recognition of this.

> These pages are designed ❗ to help you think about aspects of the investigation rather than to guide you through it step by step.

Figure 3.3 Archimedes making his discovery

Measuring the density of a liquid

The density of any substance can be worked out using the equation:

$$\text{density} = \frac{\text{mass}}{\text{volume}}$$

If mass is measured in grams (g) and volume in cubic centimetres (cm^3) then the density will be in g/cm^3.

Measuring the density of a liquid can be done by pouring the liquid into a measuring cylinder to find its volume and using a balance to find the mass. The results might look rather like those in the table.

Density of different liquids

Liquid	Mass in g	Volume in cm³
coconut oil	18.5	20
acetone	19.6	25
sea water	51.3	50

1. Explain why you would not get the mass of the liquid by just putting the measuring cylinder with the liquid on the balance and recording the reading?

2. Suggest what you could do to get the mass of the liquid.

3. Calculate the density of each of the liquids in the table.

Measuring the density of a regular solid

Solids, of course, have their own shape. If that shape is a regular one, we can calculate the volume by measuring dimensions and then using the correct equation. For example, if the solid was a cuboid, the volume would be length multiplied by width multiplied by height. The mass would be divided by this to get the density.

For example, if a 2.0 cm cube of soft rubber had a mass of 8.82 g, its volume would be 2.0 cm × 2.0 cm × 2.0 cm, which is 8.0 cm³, and the density would be 8.82 g/8.0 cm³, which is 1.1 g/cm³.

The answer can only have the same number of **significant figures** as the measurement with the least number of significant figures (in this case 8.0 – it has two significant figures).

Quantities of three materials

Material	Mass (g)	Length (cm)	Width (cm)	Height (cm)
cork	3	2.0	2.0	3.0
oak	17	2.0	3.0	4.0
tin	364	2.5	2.5	8.0

4. Look at the table above. What is the volume of the piece of cork?

5. Determine the cork's density.

6. What are the densities of the other two materials?

7. Explain why it is incorrect to use 0.708 333 33 g/cm³ as the answer for the density of oak.

Measuring the density of an irregular solid

This is, of course, the problem that Archimedes was trying to solve and his solution (apparently inspired by getting into a bath tub that was too full) was to use the idea of displacement. The solid will displace the same volume of water as its own volume. If there's room in the container, it rises. If not, it overflows.

Imagine trying to see if a gold necklace was pure. Knowing that the density of gold is 19.29 g/cm³, being able to dangle the necklace on a thread and having a glass of water full to the brim, you could carry out a simple experiment.

8. What procedure would you follow?

9. What measurements would you need to take?

10. What calculation would you then perform?

11. Why might this experiment not be very accurate?

Figure 3.4 How could you find out if this is pure gold?

Changes of state

Learning objectives:

- describe how, when substances change state, mass is conserved
- describe energy transfer in changes of state
- explain changes of state in terms of particles.

KEY WORDS

boil
changes of state
condense
conservation of mass
evaporate
freeze
melt
sublimate

When a liquid boils, the energy transferred to the liquid gives the molecules enough energy to break away from the surface.

Conservation of mass

When substances change state, mass is **conserved.** If you start with 1 kg of ice and melt it you will have 1 kg of water. Nothing has been added or removed. The process is reversible. If you freeze the 1 kg of water you will end up with 1 kg of ice again.

This is an important idea as it shows that when the state of something is altered, no material has gone away or has been added. There are the same number of particles of the substance there. It is the arrangement that has been changed, not the amount. It also shows that material has the same mass, even though it might have changed state. Boiling 1 kg of water will produce 1 kg of steam.

1 What type of change occurs when a substance changes from liquid to solid?

Changing state

Changes of state occur when substances change from one state to another. The processes of changing from one state to another are:

- **melting:** changing from solid to liquid
- **freezing:** changing from liquid to solid
- **boiling:** changing from liquid to gas at the boiling point
- **evaporating:** changing from liquid to gas when the temperature of the liquid is lower than the boiling point
- **condensing:** changing from gas to liquid
- **sublimating:** changing from solid to gas without going through the liquid state.

Changes of state are physical changes. Unlike a chemical change the change does not produce a new substance. If the change is reversed, the substance recovers its original properties.

DID YOU KNOW?

Dry ice is frozen carbon dioxide which turns directly from a solid to a gas at a temperature of −78.5 °C. The fog you see is a mixture of cold carbon dioxide gas and cold, humid air, created as the dry ice sublimates (Figure 3.5).

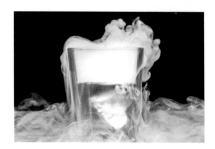

Figure 3.5 Dry ice sublimates

2 Suggest how you could prove that mass is conserved when ice melts.

3 Give an example of a change of state. Specify the material involved.

4 We often say 'it is freezing' when the temperature is cold outside. Explain why this is not an accurate statement.

Explaining changes of state

When a substance changes state, energy is transferred to change the arrangement of the particles.

When a substance melts, such as ice changing to water, energy must be transferred to the ice to change its state from solid to liquid.

In the solid state, such as ice, the particles are close-packed with strong bonds between them. Energy must be provided to weaken these bonds and allow the particles to move more freely in the liquid state (water). To change into a gas (steam) a lot of energy must be provided to break the bonds between the particles and allow the gas particles to spread out, filling the available space.

When a liquid cools down, the particles decrease their kinetic energy, allowing them to come closer together and form bonds (freezing). Energy is released to the surroundings. Energy is also released when a gas condenses to form a liquid.

Evaporation produces cooling (Figure 3.6). As a liquid warms up, the average speed of the molecules in it increases. But not all the molecules in the liquid will be travelling at the same speed (Figure 3.7). It is the faster molecules with more energy which escape from the surface of the liquid, leaving behind the slower molecules with less energy (Figure 3.8).

Figure 3.6 The sand in this pot is wet. Evaporation transfers thermal energy away from the inner pot and keeps the food cool

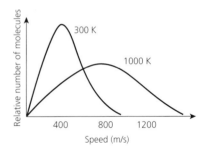

Figure 3.7 The molecules in a liquid do not all travel at the same speed

5 Explain why you hang out washing rather than leaving it in a pile to dry.

6 Explain why evaporation produces cooling.

7 Suggest why being burnt by steam is worse than being burnt by hot water.

8 Explain how sweating can help to reduce body temperature.

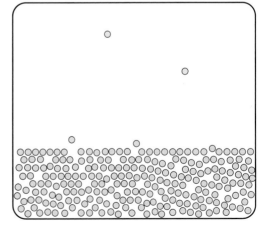

Figure 3.8 Evaporation from the surface of a liquid

Internal energy

Learning objectives:

- describe the particle model of matter
- understand what is meant by the internal energy of a system
- describe the effect of heating on the energy stored within a system.

KEY WORDS

internal energy
particle model

In air, the molecules of the gases hitting your face have an average speed of about 1600 km/h.

Particle model

Everything is made of small particles (atoms or molecules). This is the **particle model**.

The sizes of these particles are different for different materials. The particles are very hard. They cannot be squashed or stretched. But the distance between them can change. The particles are always moving. The higher the temperature the faster they move.

At the same temperature all particles have the same average kinetic energy. This means that heavy particles move slowly and light particles move quickly.

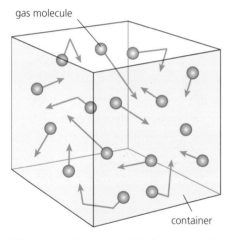

gas molecule

container

Figure 3.9 The gas particles in this container all have the same average kinetic energy

1 **Explain why the particles in solids, liquids and gases have kinetic energy**

2 **Compare the speeds of heavy particles and light particles when they are at the same temperature.**

Internal energy

The particles in solids, liquids and gases have kinetic energy because they are always moving. They also have potential energy because their motion keeps them separated. This opposes the forces trying to pull them together. The particles in gases have the most potential energy because they are furthest apart. The **internal energy** of a system is the total kinetic energy and potential energy of all the particles in the system.

3 Explain why the particles in solids, liquids and gases have potential energy.

4 Compare the amount of internal energy in a hot cup of tea with the internal energy in the Pacific Ocean. Justify your answer.

Changes in internal energy

The hotter a material is, the faster its particles move and the more kinetic energy they have. If a hot object is in contact with a cold one, energy is transferred between the two objects. When both are at the same temperature there is no further exchange of energy between them.

You can cool a glass of water by adding ice cubes to it. The faster moving molecules of water transfer energy to the ice. The internal energy store of the water decreases and the water cools down. The internal energy store of the ice increases and the ice melts.

Heating changes the energy stored in a system. It increases the energy of the particles that make it up. The increase of the internal energy store can have two effects.

When ice melts, the internal energy store of the ice increases. When there is a change of state there is no change in temperature. The increase in internal energy is used to weaken the bonds, for instance, changing a solid to a liquid.

If we carry on heating the melted ice, the internal energy store increases further, causing the temperature of the water to increase.

DID YOU KNOW?

Water is unusual in that it expands as it freezes. It is most dense at 4 °C (Figure 3.10). When a pond freezes over, there is a layer of denser water at the bottom where fish can survive.

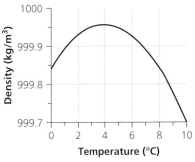

Figure 3.10 Water is most dense at 4 °C.

5 What happens to the internal energy of a system when it is heated?

6 A glass jar containing water is moved from inside a house to outside the house, where the temperature is –2 °C. Describe what happens to the internal energy of the glass and water.

7 Water can exist both in the liquid and the gaseous state at 100 °C.

  a Compare the amount of internal energy in steam at 100 °C with that of water at the same temperature.

  b Justify why using steam to power engines is more useful than using hot water.

Specific heat capacity

KEY WORDS

specific heat
capacity

Learning objectives:

- describe the effect of increasing the temperature of a system in terms of particles
- state the factors that are affected by an increase in temperature of a substance
- explain specific heat capacity.

A cup of water at 50 °C and a bath of water at 50 °C have the same temperature but the bath of water has a greater store of thermal energy.

Heating up

When a liquid is heated, the particles move faster. They gain kinetic energy and the temperature rises. The particles are close together and attract each other strongly. Their motion opposes the forces of attraction and keeps them separated. As they are moving faster, they also separate a bit more and gain potential energy. So the liquid has more kinetic energy and more potential energy. Its internal energy has increased.

1 **Describe two changes that happen to the particles when a liquid is heated.**

Increasing temperature

When a liquid is heated, its temperature increases. The temperature rise depends on:

- the mass of liquid heated
- the liquid being heated
- the energy input to the system.

2 **Dan is heating a large saucepan of water to boiling point. Explain why more energy is needed to do this than to heat a cup of water to the same temperature.**

3 **Milk does not need as much energy to raise its temperature by 10 °C as the same mass of water. Will hot milk give out less energy than the same amount of water at the same temperature when it cools down? Explain.**

Specific heat capacity

When an object is heated, energy is transferred and its temperature rises. The amount of energy needed to change the temperature of an object depends on the material the object is made from. This property is called the **specific heat capacity**.

MAKING LINKS

Chapter 1 has already introduced the idea of calculating the amount of energy in a system's energy store as its temperature changes. This is the same concept, but now we are looking at temperature change within the wider picture of internal energy and energy transfers.

The specific heat capacity of a substance is the energy needed to raise the temperature of 1 kg of the substance by 1 °C. It is given by the equation:

change in thermal energy   = mass × specific heat capacity
                           × temperature change

$\Delta E = mc\Delta\theta$

where ΔE is in J, m is in kg, c is in J/kg °C and $\Delta\theta$ is in °C.

Example: A change in thermal energy of 18 kJ of energy was supplied to a 2 kg steel block and raised its temperature from 20 °C to 40 °C. Calculate the specific heat capacity of steel.

$\Delta E = mc\Delta\theta$

$\Delta E = 18\ kJ = 18\,000\ J$

$m = 2\ kg$

$\Delta\theta = (40 - 20) = 20\ °C$

Specific heat capacity, $c = \Delta E \div (m\Delta\theta)$

$= 18\,000\ J \div (2\ kg \times 20°C) = 450\ J/kg°C$

Figure 3.11 A domestic radiator contains water which is heated by a boiler

Specific heat capacities of some materials

Material	Specific heat capacity in J/kg °C
water	4200
ice	2100
aluminium	880
copper	380

Water has a very high specific heat capacity. This means that, for a given mass, it absorbs a lot of energy when its temperature increases. It also gives out a lot of energy when it cools down (Figure 3.11).

DID YOU KNOW?

Water needs lots of energy to heat it up and gives out lots of energy when it cools down. This is why a hot-water bottle is so effective in warming a bed. An equal mass of mercury would store only 1/30th as much energy, but would take only 1/13th of the volume!

4 Calculate how much energy is needed to heat 100 g of water from 10 °C to 40 °C.

5 Water has a very high specific heat capacity. Describe a practical use of this.

6 A 2 kW electric heater supplies energy to a 0.5 kg copper kettle containing 1 kg of water.

  a Calculate the time taken to raise the temperature by 10 °C.

  b What have you assumed in doing this calculation?

7 Suggest why copper saucepans are sometimes used for cooking.

8 A 50 g copper mass is heated to 100 °C. It is then added to a cup of 50 g of water. The water has a temperature of 20 °C before the mass is added.

  a Explain what happens to the temperatures of the copper mass and the water after the mass is added.

  b Suggest whether the final temperature of the water is closer to 20 °C or 100 °C. Justify your answer.

KEY INFORMATION

Power (in watts, W) = energy (in J) / time (in s).

COMMON MISCONCEPTION

Do not confuse thermal energy transfer and temperature. Temperature is how hot an object is. A change in thermal energy does not always produce a change in temperature.

Latent heat

Learning objectives:

- explain what is meant by latent heat
- describe that when a change of state occurs it changes the energy stored but not the temperature
- perform calculations involving specific latent heat.

The latent heat of water is responsible for tornados, hurricanes and the fact that snow takes a long time to melt.

Latent heat

When you heat a solid, such as a lump of ice, its temperature rises until it starts to change to a liquid. At its melting point, the temperature stays the same until all the ice has melted (Figure 3.12).

The temperature of the liquid then rises until it starts to change into a gas. At its boiling point, the temperature stays the same until all the liquid has turned into a gas. The temperature then starts to rise again.

Latent heat is the energy needed for a substance to change a state without a change in temperature.

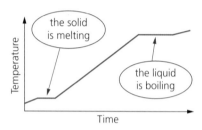

Figure 3.12 Temperature–time graph for heating a substance

1 **Describe what is happening when the graph flattens out.**

2 **Energy is needed to turn water into steam. Suggest how the particle model explains this.**

Changes of state

The amount of energy needed to change the state of a sample of a substance, without a change in temperature, depends on the mass of the sample and the type of substance. All substances have a property called **specific latent heat**.

The specific latent heat is the amount of energy needed to change the state of 1 kg of a substance without a change in temperature. Its unit is J/kg. It is given by the equation:

energy for a change of state = mass × specific latent heat

$E = mL$

where E = energy for a change of state in J

   m = mass in kg

   L = specific latent heat in J/kg.

Specific latent heat of fusion refers to a change of state from solid to liquid. **Specific latent heat of vaporisation** refers to a change of state from liquid to vapour.

Some specific latent heats

Change of state	Specific latent heat in J/kg
ice to water	340 000
water to steam	2 260 000

The specific latent heat of vaporisation is much greater than the specific latent heat of fusion. Most of this energy is used to separate the particles so they can form a gas, but some is required to push back the atmosphere as the gas forms.

3 Explain why there is no change in temperature when a block of ice melts.

4 Why is the specific latent heat of vaporisation much greater than the specific latent heat of fusion?

Latent heat calculations

(In these calculations use the specific heat capacities given in the above table.)

Many calculations of specific and latent heat require you to find the energy needed to change the state as well as the temperature of an object. This involves using both specific heat capacity and latent heat equations:

$$\Delta E = mc\Delta\theta \qquad E = mL$$

5 Calculate the energy transferred from a glass of water to just melt 100 g of ice cubes at 0 °C.

6 **a** Calculate the total energy transferred when 200 g of ice cubes at 0 °C are changed to steam at 100 °C. (Hint: Use the equation for specific heat capacity as well as the equation for specific latent heat.)

   b Sketch and label a temperature–time graph for this transfer.

7 A block of ice at −2 °C was heated. After 7304 J of energy was transferred to the ice, the ice had melted and reached a temperature of 5 °C. Calculate the mass of the ice.

DID YOU KNOW?

A jet of steam releases latent heat when it condenses (changes to a liquid). This can be used to heat drinks quickly (Figure 3.13).

Figure 3.13 Machines in cafes use steam to heat milk quickly

COMMON MISCONCEPTION

Heating does not always involve a change in temperature. When a change of state occurs, the energy supplied changes the energy stored (internal energy) but not the temperature.

Particle motion in gases

Learning objectives:

- relate the temperature of a gas to the average kinetic energy of the particles
- explain how a gas has a pressure
- explain that changing the temperature of a gas held at constant volume changes its pressure.

Inside a rocket, some of the fast-moving gas particles colliding with the walls create a force. The fast-moving gas particles moving out of the bottom of the rocket propel the rocket upwards like in Figure 3.14.

Temperature and pressure of a gas

The particles in a gas move around **randomly**. These particles have kinetic energy. They move about freely at high speed.

The higher the temperature, the faster the particles move and the more kinetic energy they have. The temperature of a gas is related to the average kinetic energy of the molecules. As the temperature increases, the particles move faster. They gain more kinetic energy.

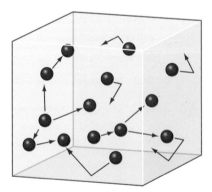

Figure 3.14 The Soyuz MS-01 spacecraft launch

① **What happens to the molecules of a gas when the gas is heated?**

② **How does the particle model explain temperature?**

The molecules of a gas collide with each other as well as with the walls of their container (Figure 3.15). When they hit a wall there is a force on the wall. Pressure is equal to the force on the wall divided by the area over which the force acts. The total force exerted by all the molecules inside the container that strike a unit area of the wall is the **gas pressure**.

③ **Describe, in terms of particles, how a gas exerts a pressure.**

④ **Suggest why the pressure inside a bicycle tyre increases when you pump more air into it.**

Changing the temperature of a gas

Air is sealed in a container (Figure 3.16). If we keep the mass and volume of the air constant, an increase in the temperature will increase the pressure of the gas.

Figure 3.15 When gas particles hit a wall of their container there is a force on the wall. This force is the pressure of the gas

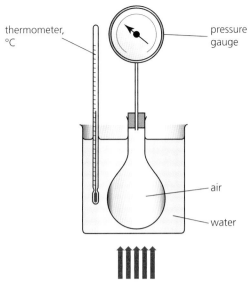

thermometer, °C

pressure gauge

air

water

Figure 3.16 Heating air in a sealed container

There are the same number of particles because the container is sealed, and there is the same mass of gas. As the container is sealed the volume of gas is also constant. As the particles have more energy, they move faster, hitting the walls more often and with greater force, increasing the pressure.

5 **How do we know that the mass of gas is constant?**

6 **Explain, using ideas about energy, why the pressure of gas increases if it gets hotter.**

Compressing or expanding gases

A gas can be compressed or expanded by pressure changes.

If the same number of particles at the same temperature are in a smaller container they hit the walls more frequently, increasing the pressure.

If the same number of particles at the same temperature are in a larger container they hit the walls less frequently, decreasing the pressure.

The pressure produces a net force at right angles to the wall of the container (or any surface).

7 **Use the particle model to explain why the particles of hot gases move very fast.**

8 **The pressure of a gas drops. Describe how this could happen.**

9 **The air in a balloon is heated. Explain why this makes the balloon expand.**

DID YOU KNOW?

The internal energy of a compressed gas can be used as an energy store. When the pressure is released in a controlled way, the expanding gas can be used to turn a turbine to generate electricity when needed.

REMEMBER!

Gas particles collide with each other and with the walls of their container, but only collisions with walls contribute to the pressure of the gas.

Increasing the pressure of a gas

KEY WORD

compressed

Learning objectives:

- describe the relationship between the pressure and volume of a gas at constant temperature
- calculate the change in the pressure or volume of a gas held at constant temperature when either the pressure or volume is increased or decreased
- explain how doing work on a gas can increase its temperature.

Pressure, volume and temperature of a gas can all change. The particle model allows you to link these variables.

Pressure, volume and temperature changes

You have already seen how if air is in a sealed container, when the temperature of a gas changes the pressure changes, and the pressure of a gas can be changed by changing the volume.

1 What happens to the pressure of a gas when the volume of its container is increased at constant temperature?

2 Suggest what would happen to the pressure of a gas at constant temperature if the volume of the gas is halved.

Linking pressure and volume, at constant temperature

Increasing the volume in which a gas is contained at constant temperature can lead to a decrease in pressure. If the volume is doubled, with the same number of particles, there will be fewer collisions between the particles and the walls of the container. The pressure will be halved.

3 Explain why the pressure is higher in the smaller box in Figure 3.17.

Decreasing the volume in which the gas is contained will result in the pressure increasing. If the volume is halved the pressure will double (Figure 3.17).

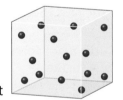

Figure 3.17 There is the same number of particles in each box

For a fixed mass of gas held at a constant temperature:

pressure × volume = constant

or pV = constant

where pressure, p, is in pascals, Pa, and volume, V, is in m³.

For a fixed mass of gas, when pressure or volume are changed:

$$p_1 \times V_1 = p_2 \times V_2$$

where p_1 is the pressure before the change, V_1 is the volume before the change, p_2 is the pressure after the change and V_2 is the volume after the change.

Example: The pressure of a gas is 1.2×10^5 Pa when its volume is 0.5 m^3. Its temperature is constant. Find its volume when the pressure changes to 2.0×10^5 Pa.

$p_1 = 1.2 \times 10^5$ Pa, $V_1 = 0.5$ m^3

$p_2 = 2.0 \times 10^5$ Pa, $V_2 = ?$

$1.2 \times 10^5 \times 0.5 = 2.0 \times 10^5 \times V_2$

$V_2 = 0.3$ m^3

3.8

4 The volume of a gas is 4 m^3 when its pressure is 200 kPa. Assuming the temperature does not change, calculate its pressure when the volume is 3 m^3.

5 The pressure of a gas is 1.8×10^5 Pa when its volume is 80 cm^3. Its temperature is constant. Calculate its volume when the pressure becomes 1.2×10^5 Pa.

MATHS

You can use cm^3 in this equation as long as you remember that both V_1 and V_2 have the same units. The same units should be used on both sides of the equation.

HIGHER TIER ONLY

Doing work on a gas

Work is the transfer of energy by a force. When a gas is **compressed**, a force is used to compress it. Energy is transferred, so work is done.

The internal energy of the gas increases, so the air molecules move faster – their kinetic energy has increased. Temperature is related to the average kinetic energy of the molecules. As the average kinetic energy of the molecules has increased, the temperature will also increase.

Work is done on air when you pump air into a bicycle tyre using a pump (Figure 3.18). The internal energy of the air in the tyre increases. In the same way, a compressor in a fridge increases the internal energy of a gas in a closed system.

Figure 3.18 Pumping up a bicycle tyre

KEY INFORMATION

Energy cannot be created or destroyed but it can be transferred from one store to another when work is done.

6 A bicycle pump is used to pump air into a bicycle tyre. Explain why doing work on a gas by using the bicycle pump increases the internal energy of the air in the bicycle tyre.

7 When using a bicycle pump, the end nearest the tyre gets hot. Explain why.

8 a Explain what happens to the internal energy and to the temperature of a gas when it expands.

   b Suggest how an expanding gas can be used in a fridge.

   c Explain where the compressor in a fridge might be used.

Google search: 'doing work on a gas'   **99**

KEY CONCEPT

Particle model and changes of state

Learning objectives:

- use the particle model to explain states of matter
- use ideas about energy and bonds to explain changes of state
- explain the relationship between temperature and energy.

KEY WORDS

bonds
fusion
internal energy
latent heat
matter
particle model
vaporisation

Everything on Earth is made of matter, but that matter isn't always the same. Matter can exist in three different states: solid, liquid, gas. The particle model of matter helps us understand the differences between those states.

Solids, liquids and gases

All matter is made up of atoms and molecules, but it is the way these atoms and molecules are held together that determines whether a substance is solid, liquid or gas. In a solid the atoms are held tightly together, forming strong regular shapes. In a liquid the atoms are held much more loosely, which is why liquids can flow. In a gas the atoms and molecules are free to move about on their own.

What determines the state of matter is the strength with which the atoms and molecules are held together by their **bonds**. All atoms also vibrate. Even in a solid, where the atoms are held together in a tight structure, every single atom is moving backwards and forwards next to its neighbour.

To get a solid to become a liquid, energy has to be added to the substance (Figure 3.19). Energy increases the vibrations of the atoms. As the vibrations increase, the bonds that hold the atoms are stretched and the atoms are pushed further apart from each other (Figure 3.20).

Adding more energy to the solid increases the vibrations even more until there comes a point where the bonds can no longer hold together. Many bonds break and the atoms are free to move – the solid has become a liquid. Adding yet more energy breaks the bonds completely and the atoms escape into the atmosphere. The liquid becomes a gas.

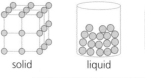

solid   liquid   gas

Add thermal energy

Figure 3.19 States of matter

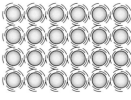

Cold

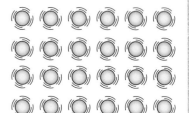

Hot

Figure 3.20 Vibrations in hot and cold solids

1 State the difference between a solid and a liquid.

2 Explain why solids expand as they are heated.

3 Explain what happens to the internal energy of a solid as it is heated.

Change of state and change in temperature

The state of matter depends on the amount of energy inside it. Heating a substance increases the amount of **internal energy** and this either raises the temperature of the system or produces a change of state. The process also works in reverse. If you take energy out of the substance it will either decrease the temperature of the system or change state from a gas back to a liquid or from a liquid to a solid (Figure 3.21).

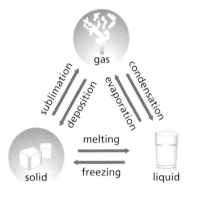

Figure 3.21 Changing between states of matter

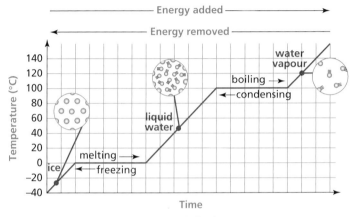

Figure 3.22 Temperature–time graph for heating water

When you heat a substance there are stages where you just get a rise in temperature and then stages where you get a change of state.

Follow the steps on the graph in Figure 3.22 which shows the three states of water: ice, liquid and water vapour.

Start with ice at −40 °C. As you heat the ice it gets warmer, but is still a solid. The slope of the graph shows the increasing temperature. Then at 0 °C energy is still being put into the ice, but there is no increase in temperature. The graph now shows a level straight line. This is where the energy is being absorbed by the substance to change it into the next state: liquid water. The amount of energy needed to change the state from solid to liquid is called the specific **latent heat** of **fusion**.

One the ice has melted into water the temperature begins to rise again. The line on the graph is sloping upwards. When the temperature reaches 100 °C we get the second change in state. Energy is absorbed but there is no rise in temperature. The amount of energy required to change the state from liquid to water vapour is called the specific latent heat of **vaporisation**.

4. Ethanol freezes at −114 °C and boils at 78 °C. Draw a temperature–time diagram for heating ethanol from −120 °C to 100 °C.

5. Explain what a line sloping down from left to right on a temperature–time graph means.

MATHS SKILLS

Drawing and interpreting graphs

Learning objectives:

- plot a graph of temperature against time, choosing a suitable scale
- draw a line of best fit (which may be a curve)
- interpret a graph of temperature against time.

A graph of temperature against time for a substance can show quite a lot about what is going on. If we know how to read a graph we can work out what that might be.

Drawing a graph

To set out the axes for a graph, the **range** of values have to be identified. The maximum and minimum values should be known.

The number of units per square on the graph paper should be chosen with care. It's essential to select a **scale** that is easy to interpret. One that uses five large squares for 100 units is much easier to interpret than a scale that has, say, three large squares for 100 units. Sometimes it is better to use a smaller scale (e.g. two large squares for 100 units) and have a scale that is easier to interpret than to try to fill the page.

A group of students conducted an experiment where they recorded the temperature of stearic acid every minute as it cooled down from 95 °C. The results are shown in the table.

Results from experiment on cooling stearic acid

Time (min)	Temperature (°C)	Time (min)	Temperature in (°C)
0:00	95.0	8:00	68.1
1:00	90.9	9:00	68.1
2:00	86.3	10:00	67.9
3:00	80.7	11:00	67.3
4:00	73.0	12:00	64.2
5:00	68.1	13:00	58.4
6:00	68.0	14:00	48.2
7:00	68.1	15:00	35.9
		16:00	25.0

1. What are the maximum and minimum values that need to be plotted?

2. Draw and label axes for plotting this dataset.

3. Plot the points and draw a line of best fit.

4. Describe the features of the line of best fit.

Interpreting a graph

The graph in Figure 3.23 shows temperature against time for heating a lump of paraffin wax. The temperature was recorded using a datalogger. The rate at which energy was supplied to the wax stayed constant. The graph shows that the temperature of the solid wax increased until about 7 minutes and then stopped rising. It then started rising again at about 15 minutes.

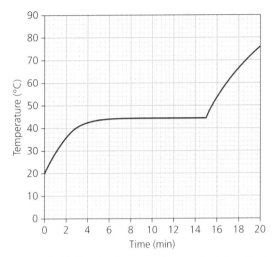

Figure 3.23 Graph of temperature against time for heating paraffin wax

5. Why is there a curve and not a straight line when the energy is being supplied at a constant rate?

6. What is happening between 7 and 15 minutes?

7. What is the melting point of paraffin wax?

Understanding the shape of the line

This graph shows a really important point. While the wax is melting, energy is being supplied but the temperature isn't changing. The energy being supplied is being used to change the state of the wax and while this is happening the temperature stays the same.

Now look at the graph you drew of temperature against time for stearic acid. This was cooling down but there are similarities.

8. Suggest an explanation for the shape of the graph.

9. What is happening between about 5 and 9 minutes?

10. What is the melting point of stearic acid?

11. Describe the shape of the graph with reference to what is happening with regard to the internal energy of the stearic acid.

12. Label the regions of the graphs for stearic acid and paraffin wax to show what is happening to the kinetic energy and the potential energy of the particles.

13. Explain how the shape of the graph in Figure 3.23 would change for a substance that had the same melting point but had a smaller specific heat capacity and a larger latent heat of fusion. Assume everything else (such as the mass of the substance and the heater) remains the same.

Check your progress

You should be able to:

Use density = mass/volume to calculate density → Use particle diagrams to communicate ideas about relative densities of different states

Use the density equation to calculate mass and volume → Link the particle model for solids, liquids and gases with density values in terms of the arrangements of the atoms or molecules

Describe changes of state as physical changes → State that mass is conserved when substances change state

Explain that changes of state are physical, not chemical, changes because the material recovers its original properties if the change is reversed → Explain how changes of state conserve mass

Describe how heating raises the temperature of a system → Describe that heating raises the temperature or changes the state of a system but not at the same time → Explain that internal energy is the total kinetic energy and potential energy of all the particles that make up a system

Describe the effect of an increase in temperature on the motion of the particles → Use the specific heat capacity equation to calculate the energy required to change the temperature of a certain mass of a substance → Use the specific heat capacity equation to calculate mass, specific heat capacity or temperature change

State that when an object changes state there is no change in temperature → Describe the latent heats of fusion and of vaporisation

Use the equation $E = mL$ → Use the particle model to explain why the latent heat of vaporisation is much larger than the latent heat of fusion

State that in the particle model the higher the temperature the faster the molecules move → Use the particle model to explain the effect on temperature of increasing the pressure of a gas at constant volume → Describe that the temperature of a gas is related to the average kinetic energy of the molecules

Recall that gases can be compressed or expanded by pressure changes → Use the equation $pV =$ constant to calculate the pressure or volume of a gas at constant temperature → Use the particle model to explain that increasing the volume of a gas, at constant temperature, can lead to a decrease in pressure

Worked example

1 **Alex was heating a beaker of water by the rays of a sunlamp. The energy transferred was 21 000 J. The time taken for the water to increase by 10 °C was 6000 seconds. Calculate the power supplied by the sunlamp.**

    3.5

This is the correct numerical answer, but

a You have not given any units to the number. Always give the unit to the quantity.

b You have not shown any working. When working out a calculation; write the equation, and show step by step how you do the calculation.

2 **Explain how raising the temperature of a gas, keeping the volume constant, increases the pressure exerted by the gas.**

The high temperature makes the molecules vibrate faster so there is a stronger force on the side of the container.

You are on the right lines. It is the force of the molecules hitting the side of the container that creates the pressure. With a gas, the molecules are no longer vibrating but they are free to move around at speed. Heating increases their kinetic energy.

3 **What does it mean to say that the changes of state are reversible?**

It means that the changes go both ways.

Correct, but you should give an example as well. Give an example of a change that goes both ways, e.g. water turns to steam and steam turns back to water.

4 **Explain what heating does to the energy stores of a system.**

Heating raises the total energy inside the system.

Yes it does, but again you should aim to give more specific information about the system. You need to mention the energy transfers that have taken place.

End of chapter questions

Getting started

1. Label these diagrams as solid, liquid and gas. | 1 Mark

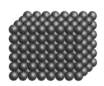

2. Describe how these models represent a solid, liquid and gas. | 2 Marks

3. Give the state that a gas turns into when it condenses. | 1 Mark

4. Give the name of the energy needed to change the state of a substance. | 1 Mark

5. Write the relationship between mass, volume and density. | 1 Mark

6. An object has a mass of 100 g and a volume of 25 cm³. Calculate its density. | 1 Mark

7. The specific latent heat of fusion of water = 340 000 J/kg. Explain carefully what this means. | 2 Marks

8. Calculate the amount of energy needed to change the temperature of 2 kg of water by 10 °C using the equation $\Delta E = mc\Delta\theta$. The specific heat capacity of water is 4200 J/kg°C. | 1 Mark

Going further

9. Write the name for the type of energy that equals the total kinetic and potential energies of the particles in a substance. | 1 Mark

10. Explain why there are two different latent heats for each substance. | 1 Marks

11. What is meant by the internal energy of a system? | 2 Marks

12. 2.0 kg of water was placed in a saucepan and heated to 100 °C. The water then completely boiled into steam.

    Here is some data about water and steam:

    density of water = 1000 kg/m³

    density of steam = 0.59 kg/m³

    latent heat of vaporisation of water = 2 260 000 J/kg

    a Explain why the density of steam is much smaller than the density of the water. | 1 Mark

    b Calculate the volume of the steam that was produced once all of the water had boiled. | 1 Mark

    c The pressure of the steam was 1.5 × 10⁵ Pa. As the steam rises up, the pressure reduces to 1.0 × 10⁵ Pa. Use your answer to part b to calculate the new volume of the steam. | 1 Mark

    d Use the equation E = mL to calculate how much energy was needed to boil the water into steam at 100 °C. | 1 Mark

13. Water in an ice cube tray is put into a freezer. Explain what happens the energy stored inside the system. | 2 Marks

More challenging

14 When a substance changes state, one energy store increases and another decreases.

   a What part of the substance is always conserved? `1 Mark`

   b State what the two possible results are when a substance is heated. `1 Mark`

15 Explain how raising the temperature of a gas, keeping the volume constant, increases the pressure exerted by the gas. `2 Marks`

16 A student wanted to measure the density of a plastic duck that floats on water. Describe an experiment the student could carry out to measure the duck's density accurately. `6 Marks`

Most demanding

17 A 2 kg block of copper is given 8.88 kJ of energy to raise its temperature by 10 °C. Calculate the specific heat capacity of copper. `2 Marks`

18 With reference to the particle model, explain how increasing the volume in which a gas is contained, at constant temperature, creates a decrease in pressure. `2 Marks`

19 A science demonstrator poured some liquid nitrogen into a plastic bottle until it was about 1/3 full. She then screwed the lid tightly on the bottle and dropped the bottle into a bucket of warm water. She then quickly ran to a safe distance before the bottle exploded.

Use ideas about energy and particles to explain the physics behind this demonstration. The boiling point of nitrogen is –196 °C. `6 Marks`

`Total: 40 Marks`

ATOMIC STRUCTURE

IDEAS YOU HAVE MET BEFORE:

AN ATOM CONTAINS PROTONS, NEUTRONS AND ELECTRONS

- The atoms of an element have the same number of protons.
- Atoms of different elements have different masses.
- All the atoms of an element have the same chemical properties.
- Atoms are unchanged in a chemical reaction.

Group																		0
1	2											3	4	5	6	7		$^{4}_{2}$He helium
$^{7}_{3}$Li lithium	$^{9}_{4}$Be beryllium											$^{11}_{5}$B boron	$^{12}_{6}$C carbon	$^{14}_{7}$N nitrogen	$^{16}_{8}$O oxygen	$^{19}_{9}$F fluorine	$^{20}_{10}$Ne neon	
$^{23}_{11}$Na sodium	$^{24}_{12}$Mg magnesium											$^{27}_{13}$Al aluminium	$^{28}_{14}$Si silicon	$^{31}_{15}$P phosphorus	$^{32}_{16}$S sulfur	$^{35}_{17}$Cl chlorine	$^{40}_{18}$Ar argon	
$^{39}_{19}$K potassium	$^{40}_{20}$Ca calcium	$^{45}_{21}$Sc scandium	$^{48}_{22}$Ti titanium	$^{51}_{23}$V vanadium	$^{52}_{24}$Cr chromium	$^{55}_{25}$Mn manganese	$^{56}_{26}$Fe iron	$^{59}_{27}$Co cobalt	$^{59}_{28}$Ni nickel	$^{64}_{29}$Cu copper	$^{65}_{30}$Zn zinc	$^{70}_{31}$Ga gallium	$^{73}_{32}$Ge germanium	$^{75}_{33}$As arsenic	$^{79}_{34}$Se selenium	$^{80}_{35}$Br bromine	$^{84}_{36}$Kr krypton	
$^{85}_{37}$Rb rubidium	$^{88}_{38}$Sr strontium	$^{89}_{39}$Y yttrium	$^{91}_{40}$Zr zirconium	$^{93}_{41}$Nb niobium	$^{96}_{42}$Mo molybdenum	$^{99}_{43}$Tc technetium	$^{101}_{44}$Ru ruthenium	$^{103}_{45}$Rh rhodium	$^{106}_{46}$Pd palladium	$^{108}_{47}$Ag silver	$^{112}_{48}$Cd cadmium	$^{115}_{49}$In indium	$^{119}_{50}$Sn tin	$^{122}_{51}$Sb antimony	$^{128}_{52}$Te tellurium	$^{127}_{53}$I iodine	$^{131}_{54}$Xe xenon	
$^{133}_{55}$Cs caesium	$^{137}_{56}$Ba barium	$^{139}_{57}$La lanthanum	$^{178}_{72}$Hf hafnium	$^{181}_{73}$Ta tantalum	$^{184}_{74}$W tungsten	$^{186}_{75}$Re rhenium	$^{190}_{76}$Os osmium	$^{192}_{77}$Ir iridium	$^{195}_{78}$Pt platinum	$^{197}_{79}$Au gold	$^{201}_{80}$Hg mercury	$^{204}_{81}$Tl thallium	$^{207}_{82}$Pb lead	$^{209}_{83}$Bi bismuth	$^{210}_{84}$Po polonium	$^{210}_{85}$At astatine	$^{222}_{86}$Rn radon	
$^{223}_{87}$Fr francium	$^{226}_{88}$Ra radium	$^{227}_{89}$Ac actinium																

$^{1}_{1}$H hydrogen

ATOMS CANNOT BE CREATED OR DESTROYED

- Atoms obey the law of conservation of matter.
- Atoms cannot be created in a chemical reaction.
- Atoms cannot be destroyed in a chemical reaction.

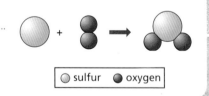

○ sulfur ● oxygen

CHEMICAL EQUATIONS SUMMARISE WHAT HAPPENS IN A CHEMICAL REACTION

- The atom is the smallest particle that can take part in a chemical reaction.
- Each element is represented by a symbol of up to 3 letters, starting with a capital letter.
- There is always the same number of atoms before and after a chemical reaction.
- There is always the same number of each type of atom before and after a chemical reaction.

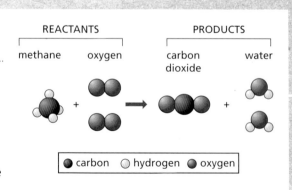

REACTANTS | PRODUCTS

methane   oxygen | carbon dioxide   water

● carbon ○ hydrogen ● oxygen

IN THIS CHAPTER YOU WILL FIND OUT ABOUT:

ARE ALL THE ATOMS IN AN ELEMENT EXACTLY THE SAME?

- The number of protons in the nucleus is called the atomic number and this defines an element.
- The number of electrons can change when an atom combines with another element chemically.
- Isotopes are atoms of the same element with different numbers of neutrons and so have a different atomic mass.
- A radioisotope has nuclei that are unstable and undergo radioactive decay.

hydrogen-1 nucleus
(1 proton)

hydrogen-2 (deuterium) nucleus
(1 proton plus 1 neutron)

IS IT POSSIBLE FOR ATOMS TO CHANGE FROM ONE ELEMENT INTO ANOTHER?

- Radioactive decay occurs when an atom emits an alpha or beta particle or a gamma ray.
- Nuclear fission occurs when a nucleus splits into the nuclei of two different atoms, emitting nuclear radiation and sometimes other particles.
- Nuclear fusion occurs when two small atomic nuclei join to form a larger nucleus.
- The nuclear radiation emitted from nuclei has many uses in medicine and elsewhere.

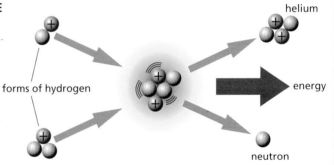

helium

forms of hydrogen

energy

neutron

CAN EQUATIONS BE USED TO REPRESENT NUCLEAR REACTIONS?

$$^{238}_{92}\text{U} \rightarrow\ ^{234}_{90}\text{Th} +\ ^{4}_{2}\text{He}$$

alpha particle

- Atoms can be represented by their atomic symbols with their atomic number and mass number.
- Subatomic particles such as alpha and beta particles are represented by symbols with their atomic number, mass number and charge.
- A nuclear equation summarises what happens in a nuclear reaction.
- There is always the same number of each type of subatomic particle before and after a nuclear reaction.

Atomic structure

Learning objectives:

- describe the structure of the atom
- use symbols to represent particles
- describe ionisation.

KEY WORDS

atomic number energy level
ionise
isotope
mass number
nucleon

Radium is one of the most powerful radioactive substances known. It emits one million times more radiation than uranium.

The structure of the atom

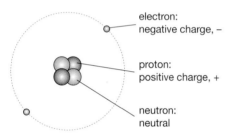

Figure 4.1 An atom contains protons, neutrons and electrons

KEY INFORMATION

The radius of an atom is about 10^{-10} m. It would take around 5 million of them placed side by side to cover the distance of 1 mm. The radius of the nucleus is less than 1/10 000th of the radius of the atom!

An atom contains protons, neutrons and electrons. The protons and neutrons are at the centre of the atom and form the atom's nucleus. Protons and neutrons are also called **nucleons**. The electrons are arranged at different distances from the nucleus (different **energy levels**).

Neutrons are neutral – they have no charge. Protons have a positive electric charge and electrons have a negative electric charge of the same size. There are the same number of electrons and protons in an atom. This means that atoms have no overall electric charge because the charges cancel out.

1. A radium atom has 88 protons and 226 neutrons. How many electrons does it have?

2. A uranium atom has 92 electrons and 238 nucleons. Determine how many protons and how many neutrons it has.

The number of protons in the atom determines what element it is. All atoms with six protons, for example, are carbon atoms. Atoms of the same element can have a different number of neutrons though. This means that elements can exist as different **isotopes**. Isotopes of an element have the same number of protons in them but different numbers of neutrons.

3. Two atoms are from the same element but they form different isotopes. Compare the numbers of electrons, protons and neutrons in the two atoms.

Using symbols to represent atoms

The number of nucleons in the atom is the **mass number** (A). The number of protons in the atom defines the element and is called the **atomic number** (Figure 4.2).

$$^{238}_{92}U \qquad ^{234}_{90}Th$$

mass number ——
atomic number ——

Figure 4.2 We represent an atom or nucleus with the element symbol, mass number and atomic number

We use the symbol representation $^A_Z X$, where:

A = mass number (or nucleon number)

Z = atomic number (or proton number)

X = chemical symbol for the element.

Z is the number of protons in the nucleus, so the number of neutrons is (A − Z).

For example, $^{14}_6 C$ has a mass number of 14 and an atomic number of 6. So it has 14 nucleons and 6 protons. The number of neutrons in the nucleus is 14 − 6 = 8 neutrons.

When an atom gains or loses electrons it becomes **ionised** – it has become charged. The atom has become an ion and can be positively or negatively charged. Ionisation can occur when ionising radiation from radioactive decay knocks one or more outer electrons out of the atoms the radiation passes through.

4. **How many protons and neutrons are in the nucleus of an atom of:**

   a $^{14}_7 N$

   b $^{235}_{92}U$?

5. **Compare the numbers of protons and neutrons in the uranium atom in Q4 and the uranium atom in Fig 4.2.**

6. **Explain how ionising radiation can turn some atoms into positive ions and some into negative ions.**

Electrons and energy levels

Electrons occupy the space around the nucleus at specific distances or **energy levels**. An individual electron can change energy levels (move closer to or further from the nucleus) only if the atom absorbs or emits electromagnetic radiation such as light for example (Figure 4.3).

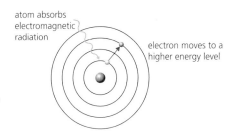

atom absorbs electromagnetic radiation

electron moves to a higher energy level

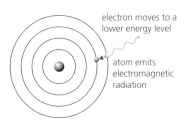

electron moves to a lower energy level

atom emits electromagnetic radiation

Figure 4.3 Absorption and emission and corresponding change in energy levels

Radioactive decay

Learning objectives:

- describe radioactive decay
- describe the types of nuclear radiation
- understand the processes of alpha decay and beta decay.

KEY WORDS

activity
alpha particle
becquerel (Bq)
beta particle
gamma ray
neutron radiation
nuclear radiation
radioisotope
random

Henri Becquerel discovered radioactivity by accident in 1896. When he left some uranium salts next to a wrapped photographic plate he found that the plate had become 'fogged'. He realised that some invisible radiation must be coming from the uranium.

Radioactive decay

Most nuclei are stable but some are not. An unstable nucleus undergoes radioactive decay to become more stable. It emits radiation as it decays. An atom with an unstable nucleus is called a **radioisotope**.

The **activity** of a radioisotope is the number of nuclear decays each second. The activity is measured in **becquerels** (Bq) or counts per second. 1 Bq = 1 count per second.

Radioactivity is a **random** process. It is not possible to predict when a nucleus will decay. If we increase or decrease the temperature of the nucleus, we still cannot predict when it will decay. Radioactive decays is independent of physical changes such as changes in temperature.

KEY INFORMATION

Lead-208 means that the mass number is 208. It is the same as $^{208}_{82}$ Pb.

1. The activity of a radioactive source is 150 Bq. How many counts would be recorded in 20 s?

2. What can you say about the nuclei of all elements larger than lead-208?

3. Explain what random means with regard to radioactive decay.

Radioactive decay produces **nuclear radiation – radiation** emitted from the nucleus. The nuclear radiation emitted may be an alpha particle, beta particle, gamma ray or a neutron. **Neutron radiation** is the release of a high-speed neutron from the nucleus, either from the nucleus of a radioactive atom or as the result of nuclear fission (see topic 4.10).

DID YOU KNOW?

The largest stable nucleus is lead-208. When the atoms from larger nuclei decay they often eventually turn into lead. This is why lead is often found near radioactive rocks.

Alpha decay

In alpha decay, an **alpha particle** is emitted from the nucleus. An alpha particle is a helium nucleus. It has 2 protons and 2 neutrons.

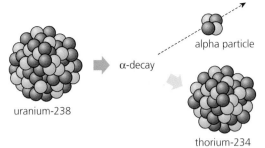

Figure 4.4 Uranium splits into thorium and an alpha particle during alpha decay

When an alpha particle is emitted from a nucleus:

- the nucleus has two fewer protons, so the atomic number (Z, the proton number) decreases by two
- the nucleus also has two fewer neutrons, so the mass number (A, the nucleon number) decreases by four
- a new element is formed.

Beta decay

In beta decay one of the neutrons in the nucleus decays into a proton and an electron. This electron is emitted from the nucleus and is called a **beta particle**.

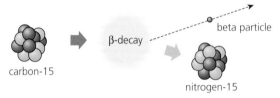

Figure 4.5 The carbon nucleus emits a beta particle and turns into nitrogen in beta decay

When a beta particle is emitted from a nucleus:

- the nucleus has one more proton, so the atomic number (Z, the proton number) increases by one
- the nucleus has one less neutron, but the mass number (A, the nucleon number) is unchanged
- a new element is formed.

4 In Figure 4.5, what are the differences between the nitrogen nucleus and the carbon nucleus?

5 What happens to a nucleus when:

a an alpha particle is emitted?

b a beta particle is emitted?

Gamma decay

In gamma decay, **gamma rays** are emitted from a nucleus. These are very high-energy electromagnetic waves. They have no charge and no mass.

The emission of a gamma ray does not cause the mass or the charge of the nucleus to change.

6 When a nucleus undergoes radioactive decay, the mass number is unchanged but the atomic number has increased by 1. What type of decay is it?

7 Suggest why gamma decay often happens immediately after alpha or beta decay.

Background radiation

Learning objectives:

- recall sources of background radiation
- describe how different types of radiation have different ionising power
- justify the selection of sources for particular applications.

KEY WORDS

background radiation
radiation dose

The radioactive decay of elements present in granite releases radon. This is a particular hazard in areas like Cornwall which are rich in granite and use it for building.

Figure 4.6 Granite houses in Cornwall

Background radiation

Background radiation is ionising radiation that is around us all the time.

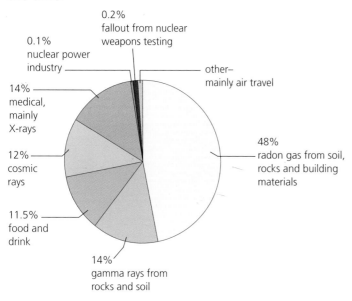

0.2% fallout from nuclear weapons testing

0.1% nuclear power industry

14% medical, mainly X-rays

12% cosmic rays

11.5% food and drink

14% gamma rays from rocks and soil

other– mainly air travel

48% radon gas from soil, rocks and building materials

Figure 4.7 The sources of background radiation

Background radiation comes from:

- natural sources such as rocks, especially granite, and cosmic rays from space
- waste products from hospitals
- waste products from nuclear power stations and other industries
- X-rays and manufactured radioisotopes used in medical procedures
- a small amount from the fallout from nuclear weapons testing and nuclear accidents.

The level of background radiation varies from place to place and from day to day. The **radiation dose** received may be

DID YOU KNOW?

Most background radiation comes from natural sources such as rocks and soil. You are naturally radioactive – you emit low levels of radioactivity from the radioisotopes in your body.

affected by your occupation and where you live but it is low level and does not cause harm.

1. Use the pie chart in Figure 4.7 to list the sources of background radiation in order, starting with the highest.

2. What percentage of background radiation comes from natural sources?

3. Suggest why background radiation varies **a** in different areas and **b** at different times.

Penetration of different types of radiation

Figure 4.8 shows the penetrating power of the three types of ionising radiation. Gamma rays are the most penetrating, able to penetrate through several metres of concrete or several centimetres of lead. Beta particles can travel a few metres in air and can be stopped by aluminium about 3 mm thick. Alpha particles are the least penetrating, and can be stopped by a few centimetres of air or a few sheets of paper.

However, when it comes to ionising power, the situation is reversed. Alpha particles have the largest mass and the highest charge, so they have the highest ionising power. They can be thought of as the radioactive equivalent to cannon balls with their huge charge and mass. If you breathe in or swallow an alpha emitter, all of the alpha particles are absorbed by your body. Beta particles do less damage and gamma rays the least because they are the least ionising.

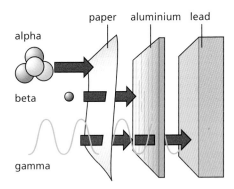

Figure 4.8 The three forms of nuclear radiation have very different penetrating powers

4. Describe the penetrating and ionising powers of beta particles.

5. Which type of radiation can penetrate a few centimetres of aluminium?

Practical applications

We can use radioactive sources for various applications where their penetrating power is useful. We might not want something that is too penetrating though and we might also want to think about the half-life. Sometimes we need something that ceases to be radioactive fairly quickly.

One example is a smoke alarm; this uses an americium source to ionise air, which can then carry a current. However, smoke prevents the ions from carrying a current and this sets the alarm off.

Another example is the monitoring of the thickness of paper in manufacture. A beta source is used and the particles detected having passed through the paper. Fewer particles mean the paper is too thick; more particles mean it's too thin.

> **REMEMBER!**
> ....
> **Alpha particles cause most ionisation.**

6. Why should the source in the smoke detector have a limited range?

7. Why should the source in the paper thickness monitoring be a beta source?

8. Radioactive tracers may be injected into the body and detected as they escape, forming images of malfunctioning organs. Why should these sources not be alpha sources?

Nuclear equations

Learning objectives:

- understand nuclear equations
- write balanced nuclear equations.

Uranium-235 undergoes a series of 14 radioactive decays to become lead-206. The series contains both alpha and beta decays.

Nuclear equations

Chemical equations show what happens in a chemical reaction. The number of atoms on each side has to be the same – the equation has to be balanced.

Nuclear equations show what happens when there are changes in the nucleus. They show the number of nucleons and charge. As with chemical equations, they have to be balanced. The number of nucleons and charge has to be the same on both sides of the equation.

1 What do chemical and nuclear equations have in common?

2 What is a key difference between chemical and nuclear equations?

Nuclear equations for alpha decay

Alpha decay and **beta decay** can be shown as nuclear equations. Figure 4.9 shows the **nuclear equation** for the alpha decay of uranium-238.

$$^{238}_{92}U \xrightarrow{\text{α-decay}} {}^{4}_{2}He + {}^{234}_{90}Th$$

Figure 4.9 The equation for the alpha decay of uranium-238

The mass numbers add up to the same number on both sides of the equation (238 = 234 + 4). This means that mass is conserved.

The atomic numbers also add up to the same on both sides of the equation (92 = 90 + 2), so the number of protons also is conserved. We can also say that the charge has been conserved.

When you write nuclear equations, make sure that the mass numbers and atomic numbers balance on both sides of the equation.

3 Copy and complete these nuclear equations for alpha decay.

a $^{226}_{\underline{}}Ra \rightarrow {}_{86}Rn + {}^{4}_{2}He$

b $^{219}_{86}Rn \rightarrow {}_{\underline{}}Po + {}_{\underline{}}He$

DID YOU KNOW?

Smoke alarms contain a small amount of americium-241 which decays by emitting an alpha particle. It becomes neptunium-237.

$^{23}_{11}$Na sodium	$^{24}_{12}$Mg magnesium											$^{27}_{13}$Al aluminium	$^{28}_{14}$Si silicon	$^{31}_{15}$P phosphorus	$^{32}_{16}$S sulfur	$^{35}_{17}$Cl chlorine	$^{40}_{18}$Ar argon
$^{39}_{19}$K potassium	$^{40}_{20}$Ca calcium	$^{45}_{21}$Sc scandium	$^{48}_{22}$Ti titanium	$^{51}_{23}$V vanadium	$^{52}_{24}$Cr chromium	$^{55}_{25}$Mn manganese	$^{56}_{26}$Fe iron	$^{59}_{27}$Co cobalt	$^{59}_{28}$Ni nickel	$^{64}_{29}$Cu copper	$^{65}_{30}$Zn zinc	$^{70}_{31}$Ga gallium	$^{73}_{32}$Ge germanium	$^{75}_{33}$As arsenic	$^{79}_{34}$Se selenium	$^{80}_{35}$Br bromine	$^{84}_{36}$Kr krypton
$^{85}_{37}$Rb rubidium	$^{88}_{38}$Sr strontium	$^{89}_{39}$Y yttrium	$^{91}_{40}$Zr zirconium	$^{93}_{41}$Nb niobium	$^{96}_{42}$Mo molybdenum	$^{99}_{43}$Tc technetium	$^{101}_{44}$Ru ruthenium	$^{103}_{45}$Rh rhodium	$^{106}_{46}$Pd palladium	$^{108}_{47}$Ag silver	$^{112}_{48}$Cd cadmium	$^{115}_{49}$In indium	$^{119}_{50}$Sn tin	$^{122}_{51}$Sb antimony	$^{128}_{52}$Te tellurium	$^{127}_{53}$I iodine	$^{131}_{54}$Xe xenon
$^{133}_{55}$Cs caesium	$^{137}_{56}$Ba barium	$^{139}_{57}$La lanthanum	$^{178}_{72}$Hf hafnium	$^{181}_{73}$Ta tantalum	$^{184}_{74}$W tungsten	$^{186}_{75}$Re rhenium	$^{190}_{76}$Os osmium	$^{192}_{77}$Ir iridium	$^{195}_{78}$Pt platinum	$^{197}_{79}$Au gold	$^{201}_{80}$Hg mercury	$^{204}_{81}$Tl thallium	$^{207}_{82}$Pb lead	$^{209}_{83}$Bi bismuth	$^{210}_{84}$Po polonium	$^{210}_{85}$At astatine	$^{222}_{86}$Rn radon
$^{223}_{87}$Fr francium	$^{226}_{88}$Ra radium	$^{227}_{89}$Ac actinium															

Figure 4.10 Part of the periodic table showing the most common isotopes

Nuclear equations for beta decay

In beta decay, a neutron changes into a proton and an electron.

$$^1_0n \rightarrow {}^1_1p + {}^0_{-1}e$$

The atomic number can also be thought of as the charge on the particle. This means that the atomic number for an electron is –1. The equation shows that the charge on both sides of the equation is 0, so charge is conserved.

Figure 4.11 demonstrates how the mass number is conserved (15 = 0 + 15) during beta decay. Figure 4.11 also shows that the charge is conserved (6 = 7 – 1) during beta decay.

4 Copy and complete these equations for beta decay.

a $^{90}_{-}$Sr $\rightarrow {}_{39}$Y + $^0_{-1}$e   b $_{15}$P $\rightarrow {}^{32}$S + $^0_{-1}$e

5 When radioactive sodium-24 decays, magnesium-24 is formed. One particle is emitted.

a Copy and complete the equation.

$_-$Na $\rightarrow {}_-$Mg + __

b What is the name of this particle?

6 This equation represents the decay of thorium-232.

$$^{232}_{90}\text{Th} \rightarrow {}^A_Z\text{X} + {}^4_2\text{He}$$

a What type of radiation is emitted?

b What are the values of A and Z?

7 Write a word equation and symbol equation for each radioactive decay:

a platinum-190, which emits an alpha particle

b rhenium-187, which emits an alpha particle

c copper-66, which emits a beta particle

d nickel-66, which emits a beta particle

e rhodium-105, which decays to palladium-105

f osmium-186, which decays to tungsten-182.

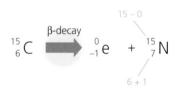

Figure 4.11 The equation for the beta decay of carbon-15

Radioactive half-life

Learning objectives:

- explain what is meant by radioactive half-life
- calculate half-life
- calculate the decline in activity after a number of half-lives.

Radioisotopes are often used to monitor a biological process in the body or in the environment. We need to know the half-lives of different radioisotopes so that we can minimise contamination or irradiation.

Half-life

We cannot predict when the nucleus of one particular atom will decay. It could be next week or not for a million years. Radioactive decay is a random process. If there is a very large number of atoms, some of them will decay each second. We plot a graph of activity against time and draw a curve of best fit. We can then use this curve to find when the activity has halved (Figure 4.12). The **half-life** of a radioisotope is the average time it takes for half the nuclei present to decay, or the time it takes for the activity to fall to half its initial level. We use half-life because we cannot predict the time it will take for all the atoms to decay.

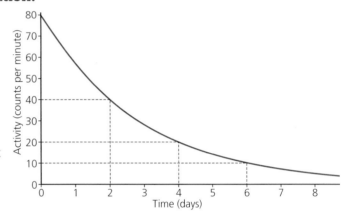

Figure 4.12 The time it takes for the activity to halve is constant

The activity of a radioactive substance gets less and less as time goes on. The graph line in Figure 4.12 gets closer and closer to the time axis but never reaches it because the activity halves each half-life.

1. Explain why you can't predict when a particular atom will decay.

2. Explain what is meant by 'half-life'.

3. What will the activity in Figure 4.12 be after:
   a 2 days? **b** 4 days? **c** 6 days? **d** 8 days?

Calculating half-life

To calculate the half-life of a radioisotope, plot a graph of count rate detected (which is proportional to the total activity) against time, as shown in Figure 4.13. The background count should be subtracted from each reading before the graph is plotted. Plot the points and then draw a smooth curve of best fit through the points. Then find several values for the half-life from the graph by finding the time for the activity in counts per minute to fall from 80 to 40, 40 to 20, 20 to 10 and so on. You should calculate the average of the values you have found. The time for the activity to halve may not be exactly the same each time. Any differences will be due to the random nature of radioactive decay.

4 Calculate the half-life of the radioisotope shown in Figure 4.13.

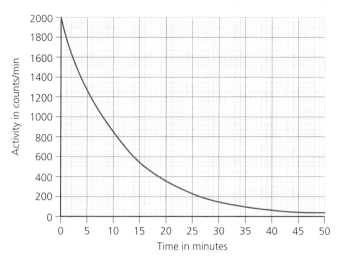

Figure 4.13

5 The table shows the activity of a radioactive sample over time.

a Draw a graph of activity against time.

b Draw a best-fit curve through the points.

c Calculate the half-life by looking at time for the activity to halve at three different points on your curve. Calculate the average.

Time in minutes	Activity in Bq
0	100
0.5	76
1.0	51
1.5	40
2.0	26
2.5	18
3.0	12
3.5	10
4.0	8

6 The activity of a radioactive sample took 4 hours to decrease from 100 Bq to 25 Bq. Calculate its half-life.

HIGHER TIER ONLY

Calculating decline in activity

For some applications such as a smoke alarm, a radioisotope with a long half-life is most suitable so that the rate of decay does not decrease significantly. Some radioisotopes are used as tracers, using the radiation they emit to trace the path of a substance the radioisope is attached to.

Radioisotopes can be used as environmental tracers (such as detecting a leak in a pipe) or medical tracers. For these applications, a short half-life is best. This means that the activity will decrease to a level similar to the background count fairly quickly because the time taken for half the radioactive nuclei to decay is very short.

DID YOU KNOW?

The half-lives of different radioactive isotopes vary from a fraction of a second to millions of years.

7 Technetium-99m is widely used in medicine. It has a half-life of 6 hours. How much of it remains after 1 day?

8 What fraction of the original sample remains after 80 minutes in Figure 4.13?

9 Caesium-134 has a half-life of 2 years. If its activity was monitored for 6 years, what fraction of the original activity level would it have dropped to after 6 years?

Hazards and uses of radiation

Learning objectives:

- describe radioactive contamination
- give examples of how radioactive tracers can be used.

Many of the servicemen who watched the first nuclear explosion in the Arizona desert then went on to develop cancers. Their bodies were contaminated by radioactivity.

Radioactive contamination

Radioactive contamination is the unwanted presence of materials containing radioactive atoms. They can be on surfaces or within solids, liquids and gases, including in the human body and on the skin.

Radioactive materials in the environment, whether natural or artificial, can expose people to risks. Radioactive materials are marked with a **hazard** symbol (Figure 4.14).

Contamination occurs when people swallow or breathe in radioactive materials. Radioactive materials can also enter the body through an open wound or be absorbed through the skin. Some radioisotopes may be absorbed by specific organs, where it is possible they could cause cancer or mutations of genes.

1 What is radioactive contamination?

2 Why is contamination a hazard?

The type and amount of radiation emitted affect the level of hazard.

The most unstable nuclei have the shortest half-lives. However, they can give out a lot of radiation in a very short time. Unstable nuclei with long half-lives may give out much smaller amounts of radiation, but this will build up over a long period of time.

How the type of radiation affects the level of contamination

The level of contamination is affected by the penetrating power of the radiation, its ionising power and the half-life of the isotope.

3 State the least hazardous form of radiation when the contamination is inside the body.

4 Explain why contamination by an alpha particle emitter is much more dangerous if it gets inside the body.

DID YOU KNOW?

If nuclear radiation enters the body it can cause serious problems if it is absorbed. The severity of the problems depends on the properties of the contamination and the half-life of the decaying substances. This is why it is important to minimise the amount of nuclear radiation getting into the body.

Radiation

Figure 4.14 Radioactive sources are marked with this hazard symbol

REMEMBER!

Alpha particles have the most ionising power but the least penetrating power. Gamma rays are the most penetrating but have the weakest ionising power.

Using medical tracers

A medical **tracer** is a radioisotope that is put into the body, either by injecting it or eating it. The tracer can be used to:

- monitor the functioning of internal organs
- check for a blockage in a patient's blood vessel (Figure 4.15).

When using radioisotope tracers, a background count should be taken several times first, in the absence of the radioisotope, and an average background count should be calculated. This value is then subtracted from readings obtained with the radioisotope.

5 **a** **Why is it important to take a background count?**

    b **Why should you take several readings of the background count?**

The tracer needs to produce nuclear radiation that can pass from inside the body to the outside so it can be detected. It also needs to be weakly ionising so that it does not do too much damage to the body.

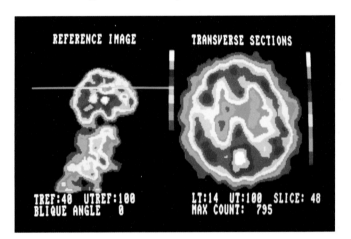

Figure 4.15 Image from a gamma tracer in a brain. The dark blue area shows where the brain tissue has died because a blood vessel is blocked

Tracers move around the body in the patient's blood. As the tracer emits radiation, we can monitor where the blood flows to. Therefore blockages in the blood flow can be detected.

Once the patient has been monitored, the tracer is no longer needed. If the tracer remained in the body it would continue to damage the cells without being of any use. Therefore it is important that we use tracers with a short half-life.

6 **A doctor has a choice of three radioisotopes that can be used as a tracer. The half-lives of the isotopes are 6 seconds, 6 hours and 6 days. Explain which isotope the doctor should use.**

7 **Suggest what type of source of nuclear radiation is the most suitable for a tracer. Explain your answer.**

8 **Suggest how a tracer could be used to check for a blockage in a patient's blood vessel.**

DID YOU KNOW?

People who have been diagnosed using medical tracers remain radioactive for several days. They have to avoid too much contact with young children and pregnant women. Nuclear radiation is particularly dangerous to bodies that are growing rapidly, such as an unborn baby.

Irradiation

Learning objectives:

- explain what is meant by irradiation
- understand the distinction between contamination and irradiation
- appreciate the importance of communication between scientists.

KEY WORDS

irradiation
mutation
peer review

Some foods are irradiated, which kills microorganisms living on them. The foods can then be kept much longer before they go off.

Irradiation

Irradiation is where an object is exposed to nuclear radiation. The exposure can originate from various sources, including natural sources and background radiation. Figure 4.16 summarises some of the ways we are irradiated each day.

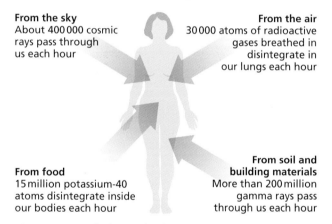

From the sky
About 400 000 cosmic rays pass through us each hour

From the air
30 000 atoms of radioactive gases breathed in disintegrate in our lungs each hour

From food
15 million potassium-40 atoms disintegrate inside our bodies each hour

From soil and building materials
More than 200 million gamma rays pass through us each hour

Figure 4.16 Each day our bodies are irradiated from different sources

DID YOU KNOW?

Low levels of irradiation have little effect on the health of humans. However, if people are exposed to high levels of irradiation it is important to consider the radiation risks for these people and their descendants.

Some things are irradiated with X-rays or gamma rays for therapeutic purposes or to sterilise them. Gamma rays are sometimes used in hospitals to sterilise food for seriously ill patients. Irradiation is also used by some supermarkets to kill bacteria on fresh food so that the food stays fresh for longer.

Nuclear radiation causes ionisation. The cells in our bodies can also be changed by radiation. DNA can be changed by nuclear radiation. This is called **mutation**. Sometimes when a cell mutates it divides in an uncontrollable way. This can lead to cancer.

1. Describe what is meant by irradiation.

2. Compare the level of irradiation we receive from food with the irradiation from the air.

Effects of irradiation

Damage to a person's cells is called damage *by* irradiation. Figure 4.17 shows the effects, which are cell death, accurate repair or misrepair causing mutation of genes.

Suitable precautions must be taken to protect against any hazard the radioactive source used in the process of irradiation may present.

3 List the three possible effects of irradiation on human body cells in order of increasing harm.

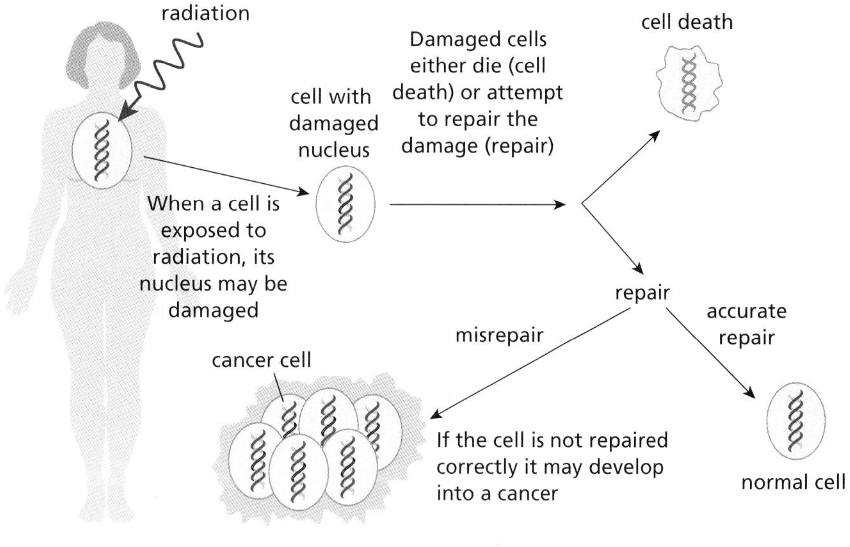

Natural and artificial sources of radiation include cosmic radiation and nuclear industry wastes respectively. Risks for people who work with nuclear radiation and become irradiated and their descendants must be considered. The calculation of these risks is used to define the tolerance levels of both an irradiation and contamination of drinking water and food products.

radiation

cell with damaged nucleus

When a cell is exposed to radiation, its nucleus may be damaged

Damaged cells either die (cell death) or attempt to repair the damage (repair)

cell death

repair

misrepair

accurate repair

cancer cell

If the cell is not repaired correctly it may develop into a cancer

normal cell

Figure 4.17 The three possible effects of irradiation on body cells

4 Why could irradiation have an effect on somebody's grandchildren?

5 Explain the difference between irradiation and contamination.

KEY INFORMATION

An irradiated object does not become radioactive.

Publishing scientific results

The first scientists to investigate nuclear radiation were unaware of its effects on their health, and many died as a result. When the first atomic bombs were exploded, scientists were not aware of the potential effects on health. They discovered that nuclear radiation can have long-term effects, causing genetic mutations which affect subsequent generations.

It is important for the findings of studies into the effects of radiation on humans to be published. This means that scientists can find out about what other scientists are doing. They can try to repeat experimental results and check the results, which is called **peer review**.

6 Explain why experimental findings should be checked by peer review.

7 Some pigeons were found to be contaminated with caesium-137, a radioisotope of caesium with a half-life of about 30 years. They were almost certainly irradiated too. Suggest why the pigeons were at greater risk from contamination than from irradiation.

Uses of radiation in medicine

KEY WORDS

radiotherapy
toxic
tumour

Learning objectives:

- compare gamma rays and X-rays
- describe some uses of nuclear radiation for medical diagnosis and therapy.

X-rays are used in a CAT (computerised axial tomography) scan to produce cross-sectional images of a part of the body. They can be used to produce a 3D image of the inside of the body.

Using X-rays and gamma radiation in medicine

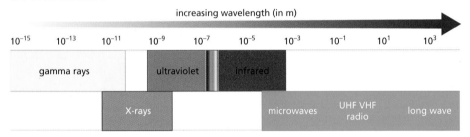

Figure 4.18 In the electromagnetic spectrum the wavelengths of gamma rays and X-rays overlap

X-rays and gamma rays are both electromagnetic waves with similar wavelengths (Figure 4.18). They are also both ionising radiation but they are produced in different ways. Gamma rays are emitted by radioisotopes, whereas X-rays are produced by X-ray machines.

X-rays and gamma rays are used to diagnose problems and treat them. They can be used to explore internal organs and bones, and to control or destroy cancerous cells (Figure 4.19).

The radioisotopes used in medicine must:

- mainly emit gamma rays
- should have a suitable half-life
- must not be **toxic** to humans.

Manufactured radioisotopes are produced with properties that make them ideal for specific uses. Sometimes radioisotopes that emit beta particles are also used.

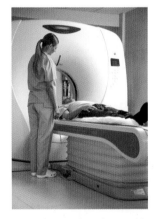

Figure 4.19 A patient undergoing radiotherapy to treat cancer

COMMON MISCONCEPTION

X-rays are not emitted from the nucleus of an atom.

1. What do X-rays and gamma rays have in common?

2. Describe the uses of manufactured radioisotopes.

Radiotherapy

Cancer cells can be destroyed by exposing the affected area of the body to extremely large amounts of radiation. This process is called **radiotherapy**. Cobalt-60 emits gamma rays and is widely used to treat cancers.

X-rays are often preferred to gamma rays because:

- X-rays are only produced when needed.
- The rate of production of X-rays can be controlled.
- The energy of the X-rays can be changed.
- You cannot change the rate of production or energy of the gamma rays emitted from a particular source.

3 Give two advantages of using X-rays rather than gamma rays.

Figure 4.20 shows how high-powered X-rays can be used in radiotherapy to destroy a **tumour** inside the body. A dose large enough to destroy the tumour would also destroy the healthy tissue it passed through. Two techniques are commonly used to protect healthy tissue:

- The source is slowly rotated around the patient with the tumour at the centre of the circle.
- The source is used in three different directions around the target area.

Both techniques minimise the side effects of the X-rays on tissues that are not cancerous. In each case, the beams intersect at the centre of the tumour.

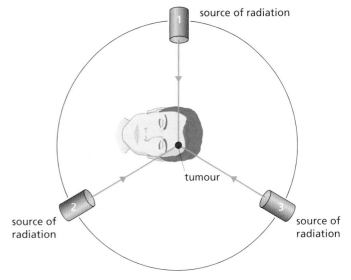

Figure 4.20 Treating a brain tumour

Brachytherapy

In brachytherapy a small sealed radioactive source, or seed, is placed in the tumour itself, to give a high dose of radiotherapy directly to the tumour but a much lower dose to the surrounding tissues. Brachytherapy is mainly used to treat cancers in the prostate gland, cervix and womb. It is sometimes given in addition to external radiotherapy.

4 How is brachytherapy different to the usual form of radiotherapy?

5 Suggest some possible problems associated with brachytherapy.

6 Explain why most radioisotopes used in brachytherapy are beta emitters.

DID YOU KNOW?

....

Iodine-123 is an artificially produced radioisotope of iodine. The thyroid gland absorbs iodine. Iodine-123, can be used to investigate problems with the thyroid gland and treat them.

Using nuclear radiation

Learning objectives:

- explore the risks and benefits of using nuclear radiation
- describe how internal organs can be explored
- understand how nuclear radiation can control or destroy unwanted tissue.

KEY WORDS

benefit
radiation dose
radiotherapy
risk

Nuclear radiation can cause cancer but it can help to cure it as well. However, treatment can only go ahead if the benefits outweigh the risks.

Making decisions

Some cancer treatments use nuclear radiation to kill cancer cells, but healthy cells get damaged as well. This leads to side effects such as:

- vomiting
- reddening and pain in the skin like sunburn
- greater risk of infection
- tiredness.

There is also a chance that the nuclear radiation can cause further cancers. Doctors and patients need to understand the **risk** of using a treatment before they decide to go ahead with it. The treatment might do more harm than good.

DID YOU KNOW?

Although cancer causes more than a quarter of all deaths in the UK, the chances of survival from cancer have doubled from 24% to 50% in the last 40 years.
[Source: Cancer Research UK]

1 **Explain why doctors decide to give some patients cancer treatment even though it has side effects.**

Some benefits of using nuclear radiation

Although using nuclear radiation has some risks, nuclear radiation can also **benefit** the health of a patient. Benefits include being able to explore internal organs and in destroying or controlling unwanted tissue.

Nuclear radiation that is used to investigate internal organs is in the form of a radioactive tracer. The tracer is often a radioisotope of an element that is used by the organ when it is working normally.

For example, the thyroid gland uses iodine. To explore whether the gland is working properly, a patient swallows a tablet with radioactive iodine in it. The thyroid gland absorbs some of this iodine and uses it just like normal iodine in the body. The radioactive iodine emits gamma rays which can be detected by a gamma camera outside the patient (Figure 4.21).

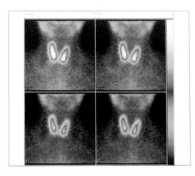

Figure 4.21 An image produced using a radioactive tracer showing two unwanted nodules in the thyroid gland. These might be treated later with nuclear radiation

2 **Hypothyroidism is a disease where the thyroid gland does not work as well as it should. Describe how doctors could use an image from a gamma camera to check whether a patient is suffering from this disease.**

Radiotherapy is where nuclear radiation is used to destroy or control unwanted tissue. For example, a patient might have radiotherapy to help remove a tumour. Radiotherapy is often used to shrink the tumour so that it is easier to remove. The remaining tumour is removed by surgery. Radiotherapy is used again on the area where the tumour was located.

3 Explain why surgery is used to remove the tumour rather than using radiotherapy to kill it completely.

4 Suggest why radiotherapy is used again, after the tumour has been removed.

Evaluating the risk

The risk of using a treatment can be evaluated by considering the **radiation dose**. Radiation dose is measured in millisieverts (mSv). It is a measure of the harm that the treatment can give you. The maximum allowed radiation dose in 1 year for workers in the nuclear industry in the UK is 20 mSv.

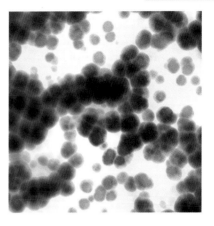

Figure 4.22 These cancer cells are too small and too spread out to be removed by surgery

5 Describe what the following quantities mean:

a activity in Bq

b dose in mSv.

> **REMEMBER!**
> ....
> In radiotherapy, a beam of radiation is rotated around the tumour. This ensures that the healthy cells surrounding the tumour only receive a small amount of radiation but the tumour receives radiation all the time.

Radiation doses from using tracers to explore internal organs. Average figures for Britain according to UK government website.

Treatment	Dose (mSv)	Equivalent time spent in background radiation
Lung ventilation scan	0.1	2 weeks
Kidney scan	1	19 weeks
Heart scan	18	7 years

6 Explain why it is more important to consider the dose given by radiotherapy rather than the activity that the patient receives.

7 Table 4.1 shows the radiation dose given to a patient when various internal organs are being explored.

a Determine which treatment is the most dangerous.

b Explain what the values in the right-hand column mean.

c Explain why doctors need to know the medical history of a patient before carrying out one of these scans.

Nuclear fission

Learning objectives:

- describe nuclear fission
- explain how a chain reaction occurs
- explain how fission is used.

KEY WORDS

chain reaction
control rods
fuel rods
graphite
moderator
nuclear fission
neutron radiation

1 kg of uranium-235 produces about the same amount of energy as 2 million kg of coal!

What is nuclear fission?

Fission is when a large and unstable nucleus, such as uranium or plutonium, splits with the release of a lot of energy. Unlike radioactive decay, spontaneous fission is rare. Usually, the unstable nucleus must first absorb a neutron (Figure 4.23). When the nucleus absorbs a neutron it splits into two daughter nuclei such as barium-141 and krypton-92. This process releases three more neutrons and lots of energy in the form of kinetic energy and gamma rays.

The release of a high-speed neutron as a product of nuclear fission is one form of **neutron radiation**. When neutron radiation is absorbed by a stable atom, it makes it unstable. This unstable atom may then undergo fission itself, or emit ionising radiation of another type.

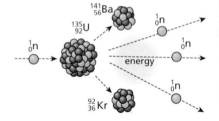

Figure 4.23 Nuclear fission

1. What is nuclear fission?
2. Describe the difference between fission and radioactive decay.

Chain reaction

The extra neutrons emitted can split more uranium nuclei causing a **chain reaction** (Figure 4.24). Each neutron released can cause the fission of another uranium nucleus if it is absorbed. A chain reaction is a whole series of fissions caused by the initial fission of one uranium nucleus. An atomic bomb involves an uncontrolled chain reaction.

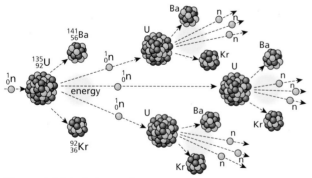

Figure 4.24 A chain reaction

3. Explain how nuclear fission can result in a chain reaction.

How a nuclear power station works

Controlled fission is used in nuclear power stations to generate electricity. Uranium is used as a fuel to heat water, making steam. The steam turns a turbine, and generates electricity (Figure 4.25).

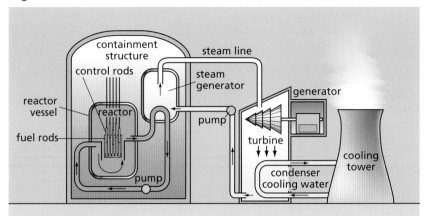

Figure 4.25 Generating electricity in a nuclear power station

4 **Compare the structure of nuclear power stations with coal-fired ones.**

Natural uranium consists of two isotopes, uranium-235 and uranium-238. The fuel used in a nuclear power station is enriched uranium, which contains a greater proportion of uranium-235 than occurs naturally.

The fission of uranium in a nuclear reactor starts the chain reaction. A chain reaction can carry on for as long as any uranium fuel remains. The chain reaction is controlled to produce a steady supply of thermal energy.

5 **How is water heated to produce steam in a nuclear power station?**

The reactor in Figure 4.26 contains a **moderator** in the form of graphite. The graphite is placed between the **fuel rods** to slow down the fast-moving neutrons emitted during fission. Slow-moving neutrons are more likely to be absorbed by other uranium nuclei, so maintaining the chain reaction.

Control rods are placed between the fuel rods in the reactor (Figure 4.26). The rods absorb neutrons so fewer neutrons are available to split more uranium nuclei. The rods can be raised or lowered to control the rate of fission.

6 **What is the purpose of a moderator in a nuclear reactor?**

7 **Describe how control rods can be used to control the temperature of a nuclear reactor.**

> **KEY INFORMATION**
>
> The products of uranium fission are radioactive. The waste also contains other materials, such as the control rods, which have been made radioactive by absorbing neutrons during the nuclear reaction.

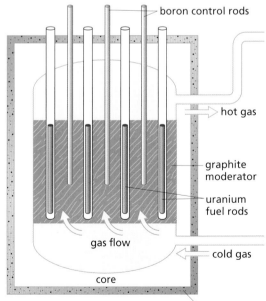

Figure 4.26 A gas-cooled nuclear reactor

Nuclear fusion

Learning objectives:

* explain nuclear fusion
* describe the conditions needed for fusion
* describe how nuclear fusion may be an attractive energy source.

The International Thermonuclear Experimental Reactor (ITER) is an international project aiming to produce more energy than it needs to create the conditions for fusion. It is scheduled to be operational in the late 2020s.

What is nuclear fusion?

Nuclear **fusion** is the joining of small light nuclei such as hydrogen and helium to form a heavier nucleus. Some of the mass of the nuclei is converted into energy (Figure 4.27).

Fusion is the energy source for stars. The fusion of two small nuclei produces huge amounts of thermal energy. In stars, hydrogen nuclei fuse at extremely high pressures and temperatures of about 10 million degrees Celsius to produce helium nuclei (Figure 4.28). The positive charges of the nuclei mean that very high temperatures and pressures are needed to bring them close enough for fusion to occur. So far we have not been able to produce these conditions on Earth so that it can be used as an energy source.

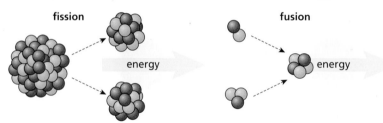

Figure 4.27 Nuclear fission and nuclear fusion

REMEMBER!

Fission is the splitting of a large nucleus such as uranium, and fusion is the joining together of two small nuclei such as hydrogen.

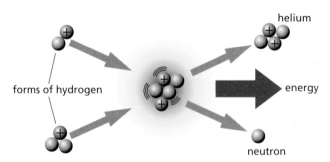

Figure 4.28 A nuclear fusion reaction that we could possibly use as an energy source in the future

1. **Compare fission and fusion.**

2. **What conditions are needed for fusion? Explain why these are needed.**

Examples of fusion reactions

Here are nuclear equations showing some of the nuclear reactions that take place in stars.

Two isotopes of hydrogen combine to form an isotope of helium:

$$^1_1H + {}^2_1H \rightarrow {}^3_2He$$

Two helium-3 nuclei then fuse to form helium-4:

$$^3_2He + {}^3_2He \rightarrow {}^4_2He + 2{}^1_1H$$

If it were possible to produce fusion on the Earth, fusion would be preferred to fission as an energy source because there is plenty of hydrogen for fusion in sea water, and waste products, mainly helium, are not radioactive.

However, it is likely that parts of the reactor would be contaminated with radioisotopes. The use of nuclear fusion for large-scale power generation is not likely for many years. If it ever does prove possible it could revolutionise energy production.

3 Why do scientists want to achieve fusion on Earth?

Figure 4.29 A hydrogen bomb exploding

Fusion bombs

The first **hydrogen** (fusion) **bomb** (Figure 4.29) was detonated in 1952. It was much more powerful than the atomic bombs dropped on Hiroshima and Nagasaki in 1945. Fusion bombs are started with a fission reaction. It creates the very high temperatures needed for fusion reactions to start.

4 Explain why scientists share details of their work with other scientists.

5 Suggest why it is easier to make a hydrogen bomb than it is to generate electricity using controlled fusion.

6 Describe some issues of producing energy using fusion in the future.

DID YOU KNOW?

In 1989 Pons and Fleischmann (Figure 4.30) reported that they had achieved fusion at room temperature (**cold fusion**). The experimental details were shared between scientists but so far no one has been able to reproduce Pons and Fleischmann's results.

Figure 4.30 Pons and Fleischmann in their lab

Developing ideas for the structure of the atom

Learning objectives:

- understand how ideas about the structure of the atom have changed
- understand how evidence is used to test and improve models.

The word atom is the Greek for indivisible. Democritus was a Greek philosopher who lived around 400 BC and developed the idea of the atom. For a long time atoms were thought to be the smallest particles.

The atom

Our theories for the structure of the **atom** have changed as new evidence has emerged.

The idea of an indivisible atom existed for over 2000 years. It was only at the start of the 20th century that scientists began to discover our current model of the atom.

At the end of the 19th century, scientists had started to question that the atom is the smallest particle. In the late 1890s, J. J. Thomson discovered the **electron** and that it is very much lighter than an atom. He proposed the **plum pudding model** of the atom where electrons are surrounded by positive charge (Figure 4.31).

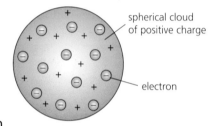

Figure 4.31 Thomson's plum pudding model

1. Does this model include protons, neutrons and electrons?

2. Are the positive and negative charges balanced in this model?

Using experimental evidence to test the model

In 1909, Hans Geiger and Ernest Marsden carried out an experiment where they aimed a beam of **alpha particles** at a thin foil of metal. According to Thomson's model, they should all pass straight through. But some of the alpha particles bounced off the foil in all directions (Figure 4.32).

This led Ernest Rutherford to propose a **nuclear model** of the atom in 1911 with a small central **nucleus** that contained most of the mass and was charged with electrons orbiting it (Figure 4.33).

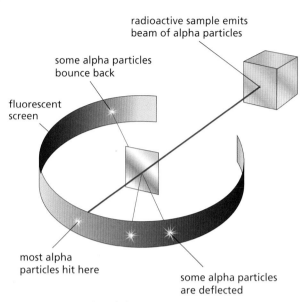

radioactive sample emits
beam of alpha particles

some alpha particles
bounce back

fluorescent
screen

most alpha
particles hit here

some alpha particles
are deflected

Figure 4.32 Results of the Geiger–Marsden experiment

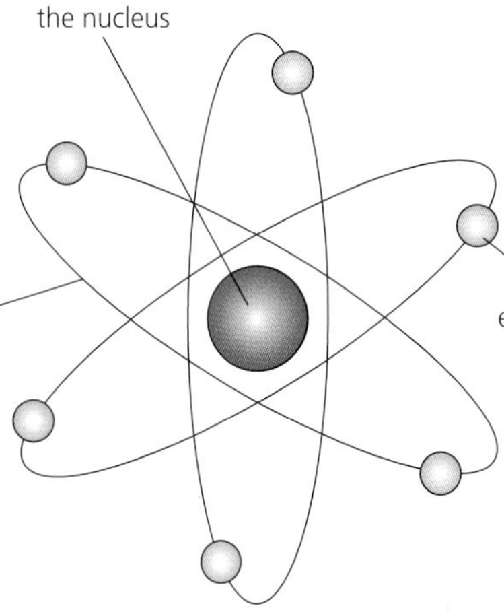

the nucleus

orbits

electrons

Figure 4.33 Rutherford's model of the atom

There was a problem with this model. Classical mechanics showed that the electrons would not stay in stable orbits and would spiral into the nucleus. So, in 1913, Niels Bohr proposed a refined model in which electrons could only be in certain orbits. This is the model we have today (Figure 4.34).

3 Explain why it was important for Geiger and Marsden to publish the results of their experiment.

4 Explain why the model of the atom had to change.

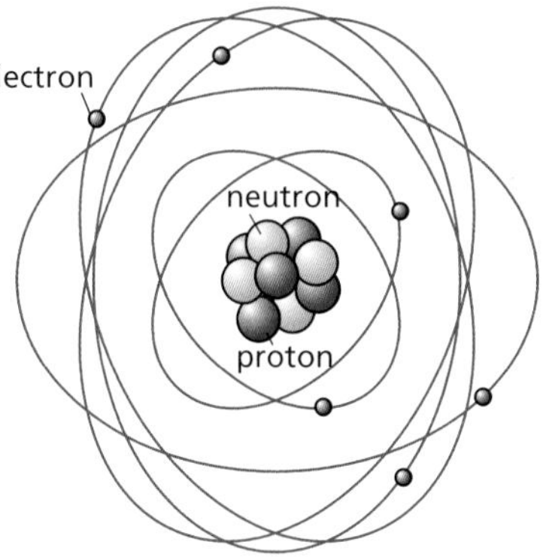

Developing the model further

The results of further experiments suggested that the positive charge of the nucleus was made up of charged particles that all had the same charge. This led to the idea of the **proton** which was discovered by Rutherford in 1920. In the same year, he suggested that the nucleus was made up of protons and neutral particles. He thought the neutral particles could be a proton and electron combined in some way. The **neutron** was then discovered in 1932 by James Chadwick.

electron

neutron

proton

Figure 4.34 Bohr's model

5 State what Rutherford's model explained.

6 Suggest how a theory becomes accepted.

MATHS SKILLS

Using ratios and proportional reasoning

Learning objectives:

- calculate radioactive half-life from a curve of best fit
- calculate the net decline in radioactivity.

When you draw a graph of activity for a radioisotope against time, the points do not fit a straight line – the line of best fit is a curve.

Measuring half-life from a graph

Working out the half-life of a radioisotope usually involves drawing a graph and using the graph to calculate the half-life.

Figure 4.35 shows the activity in counts per minute against time for a certain radioactive isotope. You can use the graph to work out the half-life of the isotope.

Look for numbers on the vertical scale that are easy to halve. For example, you can halve 40 easily to get 20.

From the 40 on the vertical axis, draw a line across to the graph. Then draw a line down to the time axis. So the activity at 2 days is 40 counts/minute.

Now repeat this for 20 counts/minute. The time is 4 days.

It takes 4 – 2 = 2 days for the activity to reduce by half from 40 to 20 counts/minute.

So the half-life is 2 days.

You should always repeat for a second pair of values, just to check. Choose another value that can be halved easily, e.g. 20. Repeating the process for halving the count rate from 20 to 10 gives the same answer, 2 days.

1　a　Calculate the half-life for the source shown in Figure 4.36.

　　b　Work out three values for the half-life.

　　c　Calculate the average of the values.

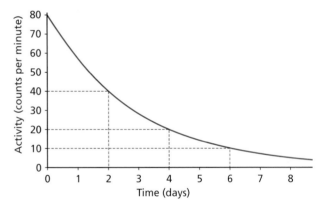

Figure 4.35 Working out the half-life of a radioisotope

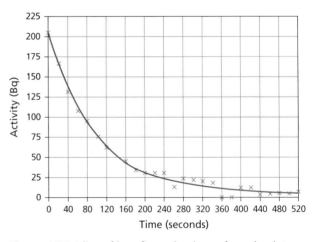

Figure 4.36 A line of best fit can be drawn from the data plotted

2 a From the data in the table plot a graph of activity against time.

Time (min)	0	10	20	30	40	50	60	70	80	90
Activity (Bq)	96	78	62	54	40	32	26	21	15	14

b Calculate the half-life.

Half-life calculations

Another way of calculating the half-life of a radioactive source is fraction reasoning. After one half-life has passed, the activity decreases by $\frac{1}{2}$. This is always true no matter when you start to measure the activity of the sample. The decrease in activity over several half-lives can be written as a sequence, as shown in the table.

Example: A radioactive source has an initial activity of 1200 counts per minute. If the count rate has fallen to 147 after 6 hours, calculate the half-life.

Starting with 1200 counts per minute at time 0:

- after 1 half-life, the activity should be 600 counts per minute
- after 2 half-lives, the activity should be 300 counts per minute
- after 3 half-lives, the activity should be 150 counts per minute. (The actual count rate is 147, showing that radioactivity is a random process.)

After 6 hours, 3 half-lives have passed, so the half-life is 2 hours.

Number of half-lives passed	Fraction of activity remaining
0	$\frac{1}{1}$
1	$\frac{1}{2}$
2	$\frac{1}{4}$
3	$\frac{1}{8}$
4	$\frac{1}{16}$
5	$\frac{1}{32}$
6	$\frac{1}{64}$

3 A radioactive isotope has an activity of 160 Bq and a half-life of 2 hours. What is the activity after 6 hours?

4 The activity of a radioactive source used to sterilise equipment in a hospital is 200 kBq. The activity is measured again in 21 years and is found to be 12.5 kBq. What is the half-life?

5 The activity of a sample decreases to $\frac{1}{16}$ th of its original value over 24 hours. Calculate the half-life of the sample.

HIGHER TIER ONLY

Net decline

In the example above, after 6 hours (3 half-lives) the activity will have reduced to $\frac{1}{8}$ th of the original. This can also be expressed as a ratio of the final value to the initial value, 1:8. This ratio is called the **net decline**.

6 Sodium-24 has a half-life of 15 hours. A sample of sodium-24 has an activity of 640 Bq.

  a Calculate the activity after 60 hours.

  b Calculate the net decline.

7 Iodine-131 has a half-life of 8 days. A sample of iodine-131 has an activity of 1800 counts per second.

  a Calculate the count rate after 32 days.

  b Calculate the net decline in the activity of the sample.

8 A sample of pure polonium-210 decays into lead-206 with a half-life of 138 days. Calculate how long it would take for the sample to contain three times as much lead as polonium.

Check your progress

You should be able to:

State that the number of protons in an element is the atomic number and the total number of protons and neutrons is the mass number → Understand that atoms of an element all have the same number of protons but can have different numbers of neutrons, giving different isotopes → Use nuclear notation to show subatomic particles in an isotope

Recognise that some isotopes called radioisotopes are unstable and decay → List some uses of radioisotopes in medicine → Describe how specific radioisotopes are used medicine

Recognise that radioisotopes have a half-life → Explain the meaning of half-life of a radioisotope → Calculate the half-life of a radioisotope

List the three types of ionising radiation resulting from nuclear activity → Describe the structure of each type of ionising radiation → Explain the properties of each type of radiation

Explain the meaning of background radiation → List different sources of background radiation → Explain why background radiation varies in different areas and in different times

Recognise the symbols used in a nuclear equation → Write nuclear equations involving alpha and beta decay → Write balanced nuclear equations for different types of nuclear reaction

Define radioactive → List the hazards of radioactive contamination → Compare and contrast irradiation and contamination

Define nuclear fission → Describe nuclear fission using a nuclear equation → Describe, using a nuclear equation, how an uncontrolled nuclear fission chain reaction can result in a huge release of energy

Define nuclear fusion → Describe nuclear fusion using a nuclear equation → Explain the potential benefits of nuclear fusion

Worked example

1 **What is an alpha particle?**

A helium atom ————————————

Not strictly correct. It is a helium *nucleus*. It would also be helpful to give more detail of its characteristics like its charge, mass, ionising power and penetrating power.

2 **Explain what you understand by the half-life of a radioisotope.**

The half-life is the time it takes to lose half its size

3 **What is the difference between irradiation and contamination?**

Contamination is where materials become radioactive. Irradiation is where an object is exposed to nuclear radiation, but it does not become radioactive itself.

A good start, but you should aim to be clearer and fuller in the explanation. The answer should refer to the activity of the radioisotope, not its size. And don't forget to mention average time.

4 **An experiment was carried out where the activity of a radioisotope was measured over time. The results are shown on the graph.**

Good answer. You could also give examples of radioactive contamination and irradiation.

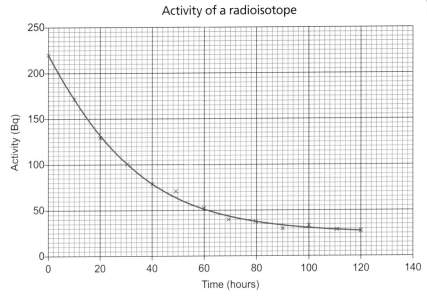

Activity of a radioisotope

a **An adjustment to the readings has not been made. Explain what it is.**

The graph has nearly levelled out above the axis. It looks like the background count has not been subtracted. ————————

This is a good answer.

b **Explain how you would work out the half-life from the graph once the adjustment had been made**

Pick a value, say 100 Bq and halve it, 50 Bq. ————————

Draw a line from 100 Bq on the vertical axis across to the graph, then down to the time axis.

Repeat for 50 Bq. Subtract the first time from the second time to get the half-life.

This gives a good answer for calculating one value of the half-life, but you should really calculate at least one more value from another part of the graph, e.g. 150 Bq and 75 Bq and calculate the average.

End of chapter questions

Getting started

1. Describe the structure of an atom using a diagram. `1 Mark`

2. Give the three types of nuclear radiation. `2 Marks`

3. Define the half-life of a radioactive element. `1 Mark`

4. Give two sources of background radiation `2 Marks`

5. Explain why background radiation varies from one place to another. `2 Marks`

6. Describe the risks and advantages of using X-rays and gamma rays in medicine. `2 Marks`

Going further

7. Define the atomic number of an element. `1 Mark`

8. Name the instrument is used to measure activity of a radioactive source. `1 Mark`

9. Explain the difference between fusion and fission. You can draw diagrams to help explain. `2 Marks`

10. Sodium can be represented by the notation $^{23}_{11}$Na.

    a What are the numbers 23 and 11 and what do they stand for? `1 Mark`

    b Describe what an isotope is. `1 Mark`

11. Iodine-131 has a half-life of 8 days. Carbon-14 has a half-life of 5715 years.

    a Explain which isotope is better suited to use as a medical tracer. `2 Marks`

    b Evaluate the risks and advantages of using radioisotopes as medical tracers. `2 Marks`

More challenging

12. What is background radiation? `1 Mark`

13. Describe the difference between irradiation and contamination. `1 Mark`

14. $^{219}_{86}$Rn decays to $^{x}_{y}$Po by emitting an alpha particle.

    Write a balanced nuclear equation for the decay. `2 Marks`

15. The activity of a radioactive sample took 60 minutes to decrease from 400 Bq to 50 Bq. Work out its half-life. `2 Marks`

16 The graph below shows an idealised graph of activity against time for a radioisotope.

    a Describe how you would expect a graph to differ from this if you measured the activity of the same radioisotope over time. `2 Marks`

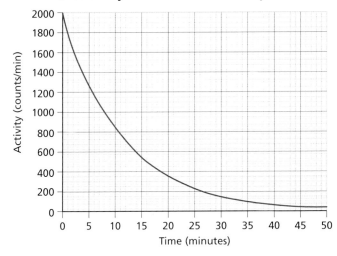

    b Calculate the half-life for this sample. `2 Marks`

Most demanding

17 Describe the characteristics of the alpha and beta particles and gamma rays. `2 Marks`

18 Explain how the evidence from Geiger and Marsden's scattering experiment led to the development of the nuclear model of the atom. `4 Marks`

19 A radioactive tracer is to be used to detect pollution in a water source.

Describe what the characteristics of the tracer should be and how you would detect it. `4 Marks`

`Total: 40 Marks`

FORCES

IDEAS YOU HAVE MET BEFORE:

SPEED AND CHANGE OF SPEED

- Average speed can be calculated by dividing the distance travelled by the time taken.
- If the speed of a car is changing it is accelerating.
- A journey can be represented on a distance–time graph.

FORCES CAUSE ACCELERATION

- Forces can speed things up or slow them down.
- The bigger the force, the bigger the change in speed.
- Forces can be contact forces, such as an engine powering a car, or non-contact forces, such as gravity.

PRESSURE

- The higher you go, the lower the atmospheric pressure is.
- The deeper you go in the ocean, the greater the water pressure becomes.
- Pressure in a solid increases when the force applied is greater and the contact area is less.

FALLING SAFELY

- Falling objects accelerate due to the force of gravity.
- Their acceleration is reduced due to the upward force of air resistance or drag.
- When a parachute opens the drag force increases.

IN THIS CHAPTER YOU WILL FIND OUT ABOUT:

HOW CAN WE DESCRIBE MOTION?

- Acceleration is a useful way of showing how speed is changing.
- When a car changes velocity it is accelerating.
- All moving objects have momentum.

HOW CAN UNDERSTANDING FORCES MAKE DRIVING SAFER?

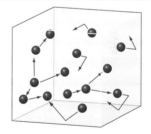

- Safety features such as seat belts and air bags reduce the force acting on the human body in an impact.
- This is because they reduce the body's rate of change of momentum.
- Reaction time is the time the brain takes to respond.
- Stopping distances depend upon the driver's reaction time and also on the vehicle and road conditions.

WHAT CAUSES PRESSURE IN A FLUID?

- Pressure in a gas is caused by particles colliding with a surface.
- We can work out the pressure in a liquid when we know how dense the liquid is and the depth.

HOW DOES THE MOTION OF A FALLING OBJECT CHANGE AS IT FALLS?

- In the absence of air resistance, all falling objects accelerate at the same rate.
- An object falling through water or the atmosphere reaches a terminal velocity when the resultant force is zero.

Forces

Learning objectives:

- describe a force
- recognise the difference between contact and non-contact forces
- state examples of scalar and vector quantities.

People sometimes use force in a wide range of ways – we talk about someone having a 'forceful personality' or the 'force of their argument'. However, in science we use force in a very particular way. We're interested in direction, size and contact.

What is a force?

A force is a push or a pull that is applied by one object on another.

Force is measured in **newtons (N)**. The force needed to pick up a bag of sugar is 20 N. The force needed to support an apple is around 1 N.

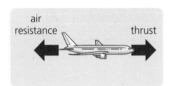

Figure 5.1 The aeroplane's engines give it a forward force or thrust. Air resistance is a frictional force that resists movement in air

Contact and non-contact forces

All forces between objects are either contact or non-contact forces.

Contact forces – the objects are physically touching, such as a parachute in contact with air (Figure 5.2).

Non-contact forces – the objects are physically separated (Figure 5.3).

Contact forces	Non-contact forces
friction	gravitational force
air resistance	electrostatic force
tension	magnetic force
normal contact force	

Figure 5.2 Air resistance – a contact force

1. Give an example of a contact force (other than the one in Figure 5.2).

2. Give an example of a non-contact force (other than the one in Figure 5.3).

Figure 5.3 Attraction and repulsion between charged objects is an example of a non-contact force

Scalar and vector quantities

A **scalar** quantity has magnitude (size) only.

A **vector** quantity has magnitude and direction.

A vector quantity can be represented by an arrow. The length of the arrow represents the magnitude and the direction of the arrow shows its direction (as in Figures 5.1 and 5.2).

Force is an example of a vector quantity.

Scalar	speed, distance
Vector	velocity, displacement

For example: speed = 10 m/s but velocity = 10 m/s due north.

Distance is a measure of how far an object moves but doesn't indicate the direction it has moved in; it is therefore a scalar quantity. **Displacement** is sometimes used instead. It includes both the distance moved and the direction so it is a vector quantity. It is measured as a straight line from the start point to the end point.

> **KEY INFORMATION**
>
> **Velocity** is speed in a particular direction.

3. a Which of the following are vector quantities?

   mass   acceleration   force

   temperature   energy   momentum

   b Explain your answers.

4. Draw two arrows to represent vectors of 20 N due south and 10 N due north.

5. A runner runs a 400 m race around a track. If he finishes at the same place as he started, what is his final distance and displacement?

Speed

Learning objectives:

- calculate speed using distance travelled divided by time taken
- calculate speed from a distance–time graph
- measure the gradient of a distance–time graph at any point.

KEY WORDS

average speed
distance–time graph
gradient
speed
tangent

The astronauts who landed on the Moon travelled at an average speed of 8000 km/h to get there.

Speed and average speed

Speed tells us how fast something is moving.

If a car has a speed of 50 km/h it means it travels 50 km in 1 hour.

$$\text{Speed } (v) = \frac{\text{distance } (s)}{\text{time } (t)} = \frac{s}{t}$$

You can convert a speed in km/h to m/s.

$$50 \text{ km/h} = \frac{50\,000 \text{ m}}{3600 \text{ s}} = 13.9 \text{ m/s}$$

We usually calculate its **average speed** because speed changes during a journey. It is rarely constant.

$$\text{Average speed} = \frac{\text{total distance}}{\text{total time}}$$

Example: A Formula One racing car driver completes a 520 km race in 2 hours. Calculate the average speed.

$$\text{Average speed} = \frac{\text{total distance}}{\text{total time}} = \frac{520 \text{ km}}{2 \text{ h}} = 260 \text{ km/h}$$

The distance travelled in a specific time can be calculated using the equation:

distance travelled = speed × time $s = v\,t$

Distance–time graphs

Drawing a graph of distance against time shows how the distance moved by a car from its starting point changes over time. It is easier to interpret than simply looking at a table of results. Look at the graphs shown in Figures 5.4a–c.

The **gradient** of a **distance–time graph** is equal to the speed of the object.

The gradient of the graph in Figure 5.4a $= \dfrac{AC}{BC}$

$$= \frac{(30 \text{ m} - 15 \text{ m})}{(6 \text{ s} - 0\text{s})}$$

$$= \frac{15 \text{ m}}{6 \text{ s}}$$

$$= 2.5 \text{ m/s}$$

So the speed is 2.5 m/s.

MATHS

To change a speed in km/h to m/s multiply by 1000/3600.
To change a speed in m/s to km/h multiply by 3600/1000.

① Explain why it is usually impossible to maintain a constant (steady) speed during a car journey.

② Abi swam 50 m in $2\frac{1}{2}$ minutes. Calculate her average speed in m/s.

③ Farah walks 10 km in $1\frac{3}{4}$ hours. What is her average speed in m/s?

④ A snail travels at an average speed of 0.5 mm/s. How far does it travel in 1 hour?

KEY INFORMATION

Typical speeds are:
- 1.5 m/s walking
- 3 m/s running
- 6 m/s cycling.

5 How would the graph change if the object went faster?

- A straight line indicates the speed is constant. A curved line shows the speed is changing.
- When the gradient increases the speed increases.
- When the gradient decreases the speed decreases.

6 Describe how the speed changes for the car in Figure 5.5.

> **KEY INFORMATION**
>
> If the speed is changing, we say the motion is non-uniform.

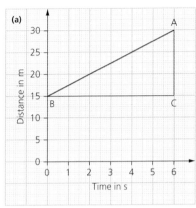
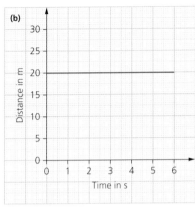
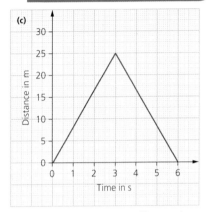

Figure 5.4 Distance–time graphs: (a) object travelling at a constant speed (b) object stationary (c) object travelling and returning to its original place

HIGHER TIER ONLY

Tangents to a distance–time graph

If an object is travelling at a steady speed, its distance–time graph will be a straight line (Figure 5.4a). However, if its speed is changing the line of the graph will be curved (Figure 5.5). We can find the speed at any particular point by drawing a **tangent** to the line and working out the gradient of the tangent.

For example, if we wanted to know the speed of this object after it had travelled 80 m, we would draw a tangent to the line at the point for 80 m and find its gradient.

The gradient of the tangent to the curve

$$= \frac{(140 \text{ m} - 20 \text{ m})}{(9 \text{ s} - 4 \text{ s})}$$

$$= \frac{120 \text{ m}}{5 \text{ s}}$$

$$= 24 \text{ m/s}$$

So the speed at this time is 24 m/s.

7 For the distance–time graph in Figure 5.6, describe how the gradient of the graph changes from 0 to 10 s. What does this tell you about how the speed changes?

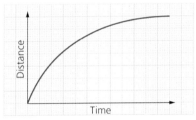

Figure 5.5 A distance–time graph for a car

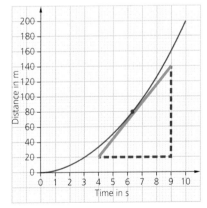

Figure 5.6 Finding the speed from a curved distance–time graph

Acceleration

Learning objectives:

- describe acceleration
- calculate acceleration
- explain motion in a circle.

• •

A car may go around a roundabout at a constant speed but it is accelerating.

Acceleration

- When the velocity of a car is increasing it is **accelerating**.
- When the velocity of a car is decreasing it is **decelerating**. Its acceleration is negative.

The faster the velocity changes the greater the acceleration. A car which has a high acceleration reaches a high speed in a short time.

1 Car A accelerates from 10 m/s to 40 m/s in 6 s. Car B accelerates from 20 m/s to 30 m/s in 5 s. Determine which car has the greater acceleration.

Figure 5.7 The Leaning Tower of Pisa

There is a story that a famous scientist called Galileo dropped two cannon balls – one large and one small – at the same time from the top of the Leaning Tower of Pisa (Figure 5.7). The balls reached the ground at the same time. This showed that both objects accelerated at the same rate and so the acceleration due to gravity does not depend on size. From this experiment, scientists deduced that all objects accelerate at the same rate under gravity (about 9.8 m/s^2).

If an object is dropped its speed increases as it falls. It accelerates because it is pulled towards the centre of the Earth due to the force of gravity.

If a ball and a feather are dropped the ball reaches the ground first because the feather has a large surface area. This slowing down force is called **air resistance** or **drag**.

DID YOU KNOW?

On the surface of the Moon the acceleration of free fall is only 1.6 m/s^2. As there is no atmosphere a feather would fall with the same acceleration as a lump of lead.

2 Explain why a ball accelerates when you drop it.

3 Why does a feather fall more slowly than a ball?

4 What happens when you throw a ball upwards?

Calculating acceleration

A sports car speeds up to 40 km/h in 5 s and a van speeds up to 40 km/h in 10 s. The acceleration of the car is twice as big as the acceleration of the van.

You can work out the acceleration of the car or van using the equation:

$$\text{acceleration} = \frac{\text{change in velocity}}{\text{time taken}}$$

$$a = \frac{\Delta v}{t}$$

If change of velocity Δv is in m/s and time t is in s, acceleration a is in m/s².

Example: A car accelerates from 10 m/s to 30 m/s in 8 s. Calculate its acceleration.

$$\text{Acceleration} = \frac{\text{change in velocity}}{\text{time taken}}$$
$$= \frac{(30 \text{ m/s} - 10 \text{ m/s})}{8 \text{ s}}$$
$$= 2.5 \text{ m/s}^2$$

> **MATHS**
>
> The symbol Δ means the difference or change in a quantity. Δv means a change in velocity.

5 A car takes 10 s to increase its velocity from 20 m/s to 40 m/s. What is its average acceleration?

6 A car has an acceleration of +2 m/s²

   a What does this tell you about the velocity of the car?

   b What is meant by an acceleration of –2 m/s²?

7 Kevin is driving his car at 24 m/s. He brakes and stops in 3 s. Find his acceleration.

8 Jane is driving at a speed of 72 km/h. She accelerates to 108 km/h in 5 s. Find her acceleration in m/s².

9 A glacier is accelerating at 4 mm/s/year. Explain carefully what this means.

> **MATHS**
>
> Remember – to change a speed in km/h to m/s: $\times \frac{1000}{3600}$.
>
> E.g. 180 km/h
> $$= \frac{(180 \times 1000)}{3600}$$
> $$= 50 \text{ m/s}$$

HIGHER TIER ONLY

Motion in a circle

When a vehicle goes round a roundabout at a constant speed its direction of movement is changing. This means that its velocity is changing because velocity is a vector. When the velocity of an object is changing, it is accelerating. There is a force towards the centre of the roundabout in the same direction as the acceleration which causes the change in velocity.

Figure 5.8 These vehicles are changing direction and so are accelerating

10 Dan and Karen are riding on a carousel at a fairground. They are moving at a constant speed. Why are they accelerating?

Velocity–time graphs

Learning objectives:

- draw velocity–time graphs
- calculate acceleration using a velocity–time graph
- calculate displacement using a velocity–time graph.

Constant acceleration does not mean constant velocity.

Velocity–time graphs

A **velocity–time graph** shows how the velocity of a moving object changes over time.

1 The table gives some velocities of a racing car at different times from the start of a race. Draw a velocity–time graph for the racing car.

> **REMEMBER!**
>
> Don't confuse distance–time and velocity–time graphs. Look at the axis labels carefully.

Time (s)	0	1	2	3	4	6	8	10	12
Velocity (m/s)	0	10	20	29	36	50	59	64	64

A graph is easier to interpret than a table of results.

When drawing velocity–time graphs remember that the velocity scale will have positive and negative values if there is a change in direction.

A **sketch graph** does not have plotted points, just the shape of the graph and the parts of the journey where the velocity is positive or negative (Figure 5.9). A sketch graph does not usually need numbers or units on the axes.

Figure 5.9a is a velocity-time graph for an object travelling at a constant velocity. Figure 5.9b shows velocity increasing steadily. Figure 5.9c shows velocity decreasing steadily to zero, then the velocity still continuing to decrease. The negative velocity shows that there is a change in direction. This could be for a car braking to a halt, then accelerating (increasing speed) in the opposite direction.

2 A tube train travels between two stations, A and B. Sketch a velocity–time graph for a journey from A to B and back to A.

Acceleration

The **gradient** of a velocity–time graph (Figure 5.10) shows how velocity changes with time. This is the **rate of change** of velocity or acceleration. A steep gradient means the acceleration is large. The object's velocity is changing rapidly.

$$\text{Acceleration} = \frac{\text{change in velocity}}{\text{time taken}} \text{ or } a = \Delta\frac{v}{t}$$

$$= \text{gradient of a velocity–time graph.}$$

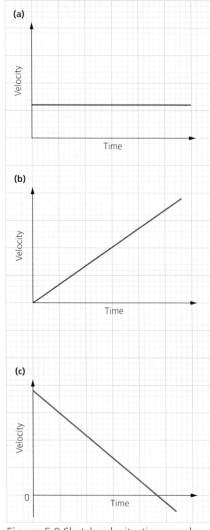

Figure 5.9 Sketch velocity–time graphs

3 Calculate the acceleration in each of the graphs in Figure 5.10.

4 A motor cycle travelling at 20 m/s takes 5 s to stop. What is its average acceleration?

5 A truck travelling at 25 m/s puts on its brakes for 4 s. This produces an acceleration of –2 m/s². What is the truck's velocity after braking?

5.4

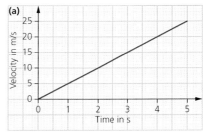

(a)

Object moving with constant acceleration from rest

HIGHER TIER ONLY

Displacement

Displacement is equal to the area under a velocity–time graph.

If we have a velocity–time graph and want to know the distance the object travelled, we can find this out by working out the area under the line.

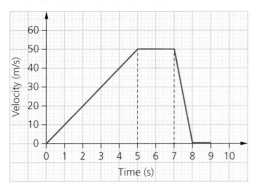

Figure 5.11

Example: For the journey shown in Figure 5.11 we can divide the area under the line into three simple shapes.

The area of the triangle on the left represents $\frac{1}{2}$ × 5 s × 50 m/s = 125 m, the area of the rectangle represents 2 s × 50 m/s = 100 m and the area of the second triangle represents $\frac{1}{2}$ × 1 s × 50 m/s = 25 m. So the total distance travelled by the object is 250 m.

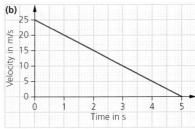

(b)

Object with constant negative acceleration slowing down to rest

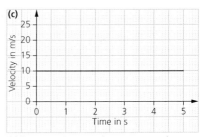

(c)

Object moving moving at a steady speed

Figure 5.10 Velocity–time graphs

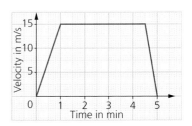

Figure 5.12 Velocity–time graph for a train

6 **a** Describe the motion of a train that has the graph shown in Figure 5.12.

   b How far did the train travel?

7 Amy is cycling through a town.

- For the first 10 s she travelled at a constant velocity.
- She started to climb a hill so she slowed down.
- At the top of the hill she had to stop for some traffic lights.
- When the lights turned green she accelerated away.
  Sketch a velocity–time graph for Amy's journey.

8 A stone was thrown upwards. Which of the graphs in Figure 5.10 could represent the motion of the stone? Describe how the graph could be extended to show the motion of the stone as it fell back down.

DID YOU KNOW?

Acceleration is a change in velocity over time. But we often use change of speed over time when calculating acceleration in situations where the direction is not important.

Calculations of motion

Learning objectives:

- describe uniform motion
- use an equation for uniform motion
- apply this equation to vertical motion.

On 15 October 1997, Thrust SSC (Figure 5.13), driven by Andy Green, broke the land speed record, reaching a speed of 763 mph, measured over a distance of 1 mile.

Describing motion

In **uniform motion** the acceleration is constant.

We use these symbols when we are describing motion:

s = displacement in m
u = initial velocity in m/s
v = final velocity in m/s
a = acceleration in m/s^2

Figure 5.13 Andy Green was required to do two 1 mile runs in opposite directions

1 What is meant by:

  a initial velocity?

  b final velocity?

2 Why are the units for acceleration different from those for velocity?

An equation for uniform motion

When an object accelerates uniformly you can use this equation to link initial and final velocities, acceleration and distance:

(final velocity)^2 – (initial velocity)^2 = 2 × acceleration × distance

$v^2 - u^2 = 2as$

Example: A car accelerates from 8.0 m/s at 2.5 m/s^2 for the next 11 m. What is its final velocity?

u = initial velocity = 8.0 m/s
a = acceleration = 2.5 m/s^2
s = displacement = 11 m

Substitute the values into the equation:

$$v^2 - 8^2 = (2 \times 2.5 \times 11)$$
$$= 64 + 55$$
$$= 119$$

Therefore $v = \sqrt{119}$
     = 11 m/s (to 2 significant figures)

Example: A train approaching a red signal has a speed of 10 m/s.

The signal then changes to green and the train accelerates. By the time it has travelled another 1000 m it is now travelling at 20 m/s. What is its acceleration?

Rearrange $v^2 - u^2 = 2as$ by dividing both sides by 2s

$$a = \frac{(v^2 - u^2)}{2s}$$

Substituting values:

$$a = \frac{(20^2 - 10^2)}{2 \times 1000}$$
$$= \frac{(400 - 100)}{2000}$$
$$= \frac{300}{2000} = 0.15 \, \text{m/s}^2$$

3 An aircraft accelerates from rest at 2.5 m/s². Its take-off speed is 60 m/s. What length of runway does it need to take off?

4 A car accelerates from rest at 3 m/s² along a straight road. How far has the car travelled after 4 s?

5 A train is travelling at 40 m/s when its brakes are applied. This produces a deceleration of 2 m/s². Determine the distance the train travels before stopping.

> **REMEMBER!**
> ....
> When the object is slowing down, v will have a smaller value than u.

Applying the equation to vertical motion

Useful points to remember:

- If there is no air resistance, gravity gives a falling object an acceleration of approximately 9.8 m/s² downwards. When a ball is thrown upwards it decelerates. Acceleration is then equal to –9.8 m/s². Acceleration is negative.
- When the ball falls back to the ground it accelerates. Acceleration = +9.8 m/s².
- The equation can only be used when an object travels with constant uniform acceleration in a straight line.

Example: A ball is thrown vertically upwards at 16 m/s. Ignoring air resistance and taking $g = 9.8$ m/s² calculate how high it goes.

We know that when it reaches the highest point it will be (momentarily) stationary, so velocity, $v = 0$.

$u = 16$ m/s
$v = 0$ m/s
$a = -9.8$ m/s²
$v^2 - u^2 = 2as$
$0 - (16)^2 = (2 \times -9.8 \times s)$
$0 - 256 = -19.6s$
$19.6s = 256$, so $s = 13$ m (to 2 significant figures)

6 A ball is thrown upwards with a velocity of 11 m/s. How high does it go? Assume $g = 9.8$ m/s².

7 An apple drops from a tree and falls 2.5 m to the ground. What is its speed when it hits the ground? Assume $g = 9.8$ m/s² and ignore air resistance.

Heavy or massive?

Learning objectives:

- identify the correct units for mass and weight
- explain the difference between mass and weight
- understand how weight is an effect of gravitational fields.

KEY WORDS

gravitational field strength
mass
newtonmeter
weight

Everyone knows that most things fall to the ground and that it's gravity that causes this to happen. This doesn't mean that gravity only acts downwards. It's a force of attraction between any two objects, such as you and the person you're sitting next to. However, you only notice it if at least one of the objects is massive.

Weight and mass?

Some people say they want to lose weight and they may keep weighing themselves. The scales are probably marked in kilograms (kg). Even if you never use scales, you'll know that you buy food such as sugar and flour by weight. The bags are marked in kilograms. In these examples weight is being confused with mass.

It's important, in science, to understand that mass and weight are not the same thing.

Mass is the amount of substance that is present in an object. It is measured in kilograms. **Weight** is the force acting on that mass, if it is in a gravitational field. As weight is a force; it is measured in newtons (N).

1 **An astronaut on the Moon feels much lighter than on Earth. Have they lost weight?**

2 **Denzil says that a weighing scale is actually a forcemeter. Is he right? Explain.**

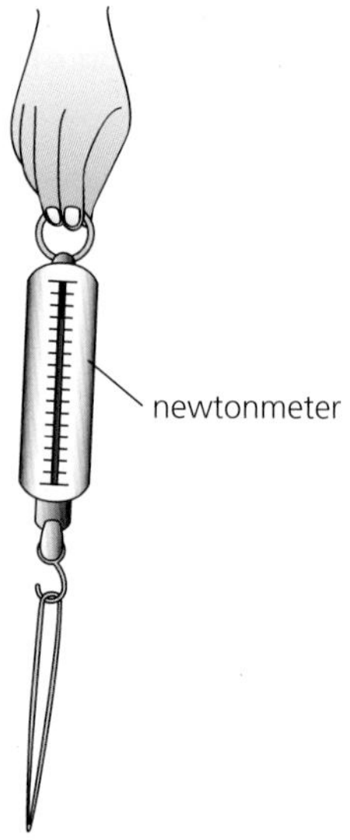

Figure 5.14 The weight of an object can be measured using a **newtonmeter**

Weight is a force

Gravity is a non-contact force. It is useful to think of it as a force field; anything in a gravitational field will experience a force of attraction. We are all in the Earth's gravitational field; it is attracting us towards the Earth.

Weight is the force acting on an object due to gravity. The weight of an object depends on the **gravitational field strength** at the point where the object is, and on the mass of the object. The weight of an object can be calculated using the equation:

weight = mass × gravitational field strength

$$W = mg$$

KEY INFORMATION

Remember that in science we measure mass in kilograms and weight in newtons. We use different units because they are different things, but there is a relationship between them.

Weight, W, is measured in newtons, N (Figure 5.14), mass, m, in kg and gravitational field strength, g, in N/kg. On Earth g is taken to be 9.8 N/kg. g is also given as 9.8 m/s^2. The units are equivalent.

Example: Calculate the weight on Earth of a 5.0 kg mass. Assume g = 9.8 N/kg.

$W = mg$

    = 5.0 kg × 9.8 N/kg

    = 49 N

3 Write down the weight on Earth of objects having a mass of:

  a 7 kg

  b 0.5 kg

  c 400 g

4 A steel block weighs 30 N on Earth.

  a What is its mass?

  b The steel block is taken to Mars where it weighs 11.1 N. Calculate the gravitational field strength on Mars.

Losing weight in space

Mass is a property of an object and is constant wherever you are. Weight, however, depends not only on the mass but also on the gravitational field strength. On the Moon, gravity (at the surface) is around a sixth of that on Earth, so everything weighs one sixth of what it does on Earth. Scales are forcemeters where the scale is given in kilograms. Standing on the scales in a lunar space station would give a reading one sixth of that on the Earth. Your mass is a constant but your weight is much less.

In deep space, well away from any stars or planets, you would be weightless. You would still have mass though, and if you wanted to move from one end of your spacecraft to the other, you would still need to apply a force.

5 Alex says that weighing scales should all carry a label saying 'calibrated for use on planet Earth only'. Is she right?

6 An astronaut in deep space (where there is no gravity) is at one end of her craft and wants to move to the other. She pushes against the inside of the craft with a force of 25 N and this causes her to accelerate at 0.5 m/s^2. What is her weight?

7 Describe a measuring instrument that could give a correct measurement of mass on the Earth and on the Moon without having to change any settings.

KEY INFORMATION

The weight of an object can be considered to act at a single point. We call this point the object's centre of mass.

Figure 5.15 The weight is shown as a force vector passing through the object's centre of mass, X

Forces and motion

Learning objectives:

- understand what a force does
- explain what happens to an object if all the forces acting on it cancel each other out
- analyse how this applies to everyday situations.

ABS (anti-lock braking system) brakes stop a car more quickly by rapidly pumping the brakes and preventing skidding.

Thinking about forces

If we think about an object, we can usually identify the forces that are acting on it. For example, an apple growing on a tree (Figure 5.16) has the force of gravity acting downwards; this is its weight. There is another force acting, vertically upwards, through the stalk. These two forces are equal and opposite. They cancel out and the apple doesn't move.

Let's think of another situation: that of a parachutist. The parachutist jumped out of an aircraft and is falling. After a while, the parachutist reaches a steady speed called the terminal velocity. There is a downwards force weight, and an upwards force, which is called drag, or air resistance (Figure 5.17). When these two forces are equal and opposite, they cancel out.

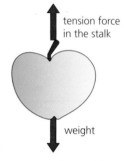

Figure 5.16 **Balanced forces** on an apple hanging from a stalk

Figure 5.17 Balanced forces on a parachutist falling at a steady speed

1. The weight of the apple in Figure 5.16 is 1 N. What is the force upwards in the stalk?

2. When the parachutist in Figure 5.17 opens the parachute:

   a What will happen to the size of the drag force?

   b Will the two forces still be in balance immediately after the parachute opens?

3. Is it *always* true that when all the forces on an object balance out it will be stationary?

Newton's first law

DID YOU KNOW?

For a parachutist in free fall, before the parachute is released, the terminal velocity is around 50 m/s (180 km/h). However, by assuming a diving position this can increase to twice as much.

resistive forces

forward force from engine

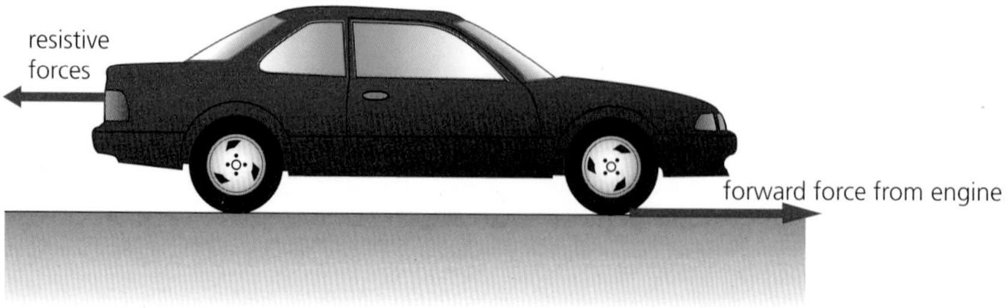

Figure 5.18 If the forces cancel out the resultant force is zero

An object may have several forces acting on it (Figure 5.18). A number of forces acting on an object may be replaced by a single force that has the same effect as all the original forces acting together. This single force is called the **resultant force**. If the forces are in balance they cancel each other out and the resultant force is zero. The object behaves as if there is no force on it at all. The object will be in **equilibrium**; it will not accelerate.

Newton's first law says that, if the resultant force acting on an object is zero, it will:

- if stationary, remain stationary
- if moving, keep moving at a steady speed in a straight line.

4 If the resultant of two forces is zero, what must be true about:

  a their size?

  b their direction?

5 Look at the two forces shown in Figure 5.18. If the resultant of the two forces is zero, what must be true about the sizes of those forces?

6 In addition to the forces shown in Figure 5.18, there will also be weight, acting downwards, and a reaction force, acting upwards. How do these also obey Newton's first law?

Applying Newton's first law

One of the most important questions to ask about a situation in which forces are acting on an object is whether the resultant force in a certain direction is zero. In this case Newton's first law applies and a stationary object will remain at rest and if the object is moving it will continue to move in a straight line and at constant velocity. Examples of this are a boulder resting on the ground and a bicycle being pedalled along a level road at steady speed in a straight line.

7 Explain why there is a zero resultant force for a boulder on the ground.

8 Explain why there is a zero resultant force in the horizontal direction for a bicycle being pedalled along a level road at steady speed in a straight line.

9 Explain why, in the example of the bicycle, if the cyclist gets tired, the resultant force may no longer be zero.

10 Describe the motion of the Earth if the Sun suddenly vanished.

DID YOU KNOW?

Deep in space, with no drag or gravitational forces to affect it, a spacecraft moving with its rockets off will keep moving forever without slowing down or speeding up.

REMEMBER!

A normal contact force is exerted on an object by, for example, the surface of a floor or wall it is in contact with. The force is at right angles to the surface. In physics, 'normal to' means 'at right angles to'.

Resultant forces

Learning objectives:

- calculate the resultant from opposing forces
- draw free-body diagrams to find resultant forces
- understand that a force can be resolved into two components acting at right angles to each other.

There's always more than one force acting in a situation. For example, when you pull a box across the floor the friction force acts in the opposite direction to the pulling force on the box.

Combining forces

When there are several forces acting on an object, we can work out what their combined effect is. If one child is behind a sledge, pushing it along, and another is in front, pulling it, we can combine these forces by adding them. In a tug-of-war, two teams are pulling in opposite directions. The force applied by one team should be subtracted from that applied by the other team to find out what the combined force is. The combined force is called the **resultant force**.

1. Gemma and Alan are pushing a packing case across the floor, each applying a force of 40 N. The friction opposing the motion is 50 N. What is the resultant force?

2. Jo is driving her car along a level road. The engine is providing 1500 N of force. Air resistance is opposing the motion with a value of 1000 N and friction accounts for another 500 N. What is the resultant force?

3 N   3 N

Figure 5.19 The forces are **balanced**. The resultant force is zero

6000 N   2000 N

Figure 5.20 The forces are **unbalanced**. The resultant force is not zero

HIGHER TIER ONLY

Free-body diagrams

A **free-body diagram** shows the magnitude and direction of the forces acting on an object. This saves us from having to spend a lot of time drawing the exact shape. The object (such as the aircraft shown in Figure 5.21) is represented by a point. The force arrows all start from the centre of the point.

A free-body diagram for the jet aircraft is shown in Figure 5.22. It shows all the forces acting. The direction of the forces is shown by the direction of the arrows and the size of the forces by the length of the arrows.

3. Draw a free-body diagram for a stationary car.

4. Draw a free-body diagram for an aeroplane flying horizontally with positive acceleration.

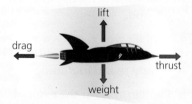

Figure 5.21 Forces on an aircraft

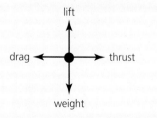

Figure 5.22 The free-body diagram for the jet aeroplane

Finding forces from a vector diagram

You can determine the magnitude and direction of a resultant force by drawing a scale diagram.

To work out the resultant of the forces in Figure 5.23, draw the 3 N force as shown in Figure 5.24. Then draw the 4 N force with its tail joining on to the head of the 3 N force. The lengths of the arrows represent the magnitudes of the forces. Choose a suitable scale, such as 1 cm = 1 N.

The resultant force is the imaginary line that goes from A to B in Figure 5.24. The magnitude of the resultant force is found by measuring the length of the line and the direction of the resultant force is found by measuring the angle θ. In this case you should find that the resultant force is 5 N at an angle of 37° above the horizontal.

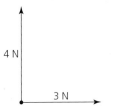

Figure 5.23 Free-body diagram for two forces acting at right angles

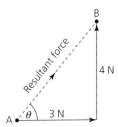

Figure 5.24 A scale diagram to calculate the resultant force

5. **Two forces act on an object. One force is 5 N to the right and the other force is 12 N upwards. Draw a scale diagram to find the magnitude and direction of the resultant force.**

6. **A 12 N force is pulling an object to the right and a 16 N force is pulling it downwards. A third force is needed to keep the object still, so that there is no resultant force. Determine the size of the third force by drawing a scale diagram.**

Looking at this the other way round, if you take a single force you can find two forces which, when combined, will produce a resultant force with the same magnitude and direction as the original. This is called **resolving the force** into **components**. It is useful to resolve a force into two perpendicular components, such as the horizontal and vertical components of the force. For example the 5 N force in Figure 5.24 has the same effect as the 3 N and 4 N forces acting together at right angles. You can use a scale diagram to determine the size of the components.

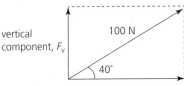

Figure 5.25 Scale diagram for example

Example: A force of magnitude 100 N acts in a direction of 40° to the horizontal. Resolve this force into horizontal and vertical components.

Draw a scale diagram with the 100 N force at 40° to the horizontal. Choose a scale that will give a large triangle. At 5 N:1 cm the force arrow is 20 cm long.

Draw a straight line down to the *x*-axis and measure the length of the line. This is the horizontal component of the force.

Length = 15.4 cm so horizontal component = 77 N

Draw a straight line across to the *y*-axis and measure the length of the line. This is the vertical component of the force.

Length = 12.8 cm so vertical component = 64 N

7. **A 14.1 N force acts at 45° above the horizontal to the right. Determine the vertical and horizontal components of this force.**

8. **The 14.1 N force in question 7 acts together with a 14.1 N force which points at 45° below the horizontal. Determine the resultant force produced by these two forces.**

Forces and acceleration

Learning objectives:

- explain what happens to the motion of an object when the resultant force is not zero
- analyse situations in which a non-zero resultant force is acting
- explain what inertia is.

The record for the fastest object made by humans is held by the Helios 2 spacecraft, which reached over 246 000 km/h. It has a very elliptical orbit around the Sun, which means it accelerates as a result of the Sun's massive gravitational field.

What does a force do?

When the driver presses on the accelerator pedal, it increases the forward **force** of the engine. This makes the car accelerate.

The force of the engine acts forwards and there will be other forces opposing this: friction and air resistance. However, at this point the force from the engine is greater than the opposing forces and so the car accelerates (Figure 5.26). It speeds up.

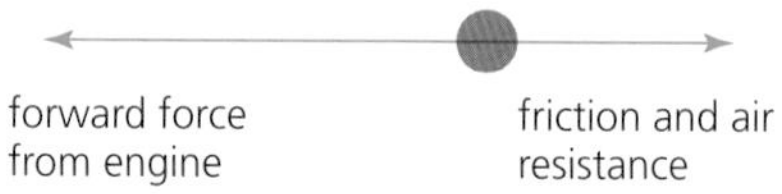

forward force from engine — friction and air resistance

Figure 5.26 The car accelerates

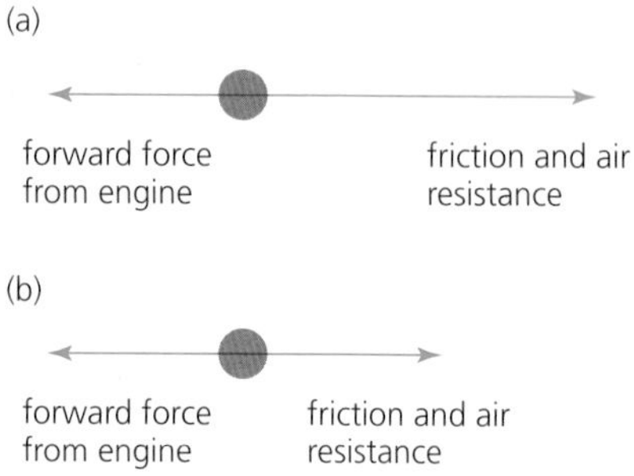

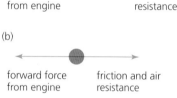

(a)

forward force from engine — friction and air resistance

(b)

forward force from engine — friction and air resistance

Figure 5.27 Forces on a car

1 In which direction are the resultant forces acting in Figures 5.26 and 5.27?

2 The driver of the car in Figure 5.27(b) presses the accelerator harder so that the speed increases. What will happen to the resultant force?

Newton's second law

Force, mass and acceleration are linked by the equation:

resultant force = mass × acceleration

$F = ma$

where F is the **resultant force** in N, m is the mass in kg and a is the acceleration in m/s². This is **Newton's second law**.

The resultant force is a single force that has the same effect as all the original forces acting together.

Example: A car has a mass of 1000 kg. What force is needed to give it an acceleration of 5 m/s²?

$F = ma$

$= 1000 \text{ kg} \times 5 \text{ m/s}^2$

$= 5000 \text{ N}$

3 A car of mass 1200 kg has a resultant forward force acting on it of 4200 N. Calculate its acceleration.

4 The weight of an apple is 1 N and it accelerates downwards at 10 m/s². What is its mass?

5 What force is needed to accelerate a 4000 kg rocket upwards at 2 m/s²? ($g = 9.8\,\text{N/kg}$)

HIGHER TIER ONLY

Inertia

Figure 5.28 Once a massive tanker is moving it is difficult to stop

Massive objects are hard to start to move and also very difficult to stop (Figure 5.28). They have an inbuilt reluctance to start or stop moving. This is called **inertia** (from the Latin for laziness). Inertia is the natural tendency of objects to resist changes in their velocity.

Inertial mass is a measure of how difficult it is to change the velocity of an object.

Inertial mass is defined by the ratio of force over acceleration:

$$\text{inertial mass} = \frac{\text{force}}{\text{acceleration}}$$

DID YOU KNOW?

Some large ships can take 10 km to stop from a cruising speed of 12 km/h.

6 What is inertia and how do you calculate inertial mass?

7 Another form of mass is **gravitational mass**. Suggest what the difference is between gravitational mass and inertial mass.

8 An object has a mass of 2 kg. What are its inertial mass and its gravitational mass?

REQUIRED PRACTICAL

Investigating the acceleration of an object

Learning objectives:

- plan an investigation to explore an idea
- analyse results to identify patterns and draw conclusions
- compare results with scientific theory.

KEY WORDS

direct proportion
inverse
    proportion
Newton's second
    law

There is a very important relationship between force, mass and acceleration. This is a fundamental idea in Physics, yet it is possible to demonstrate it using fairly straightforward apparatus and analysing the results with care.

Planning your investigation

The relationship being explored here is the one between force, mass and acceleration. The acceleration can be determined in several ways; for example, light gates and a data logger or a ticker-timer could be used. The effects of varying either force or acceleration can then be investigated.

Figure 5.29 A smartphone accelerometer chip measures acceleration. This enables the auto-rotate function to turn the screen display vertical or horizontal when the phone is moved

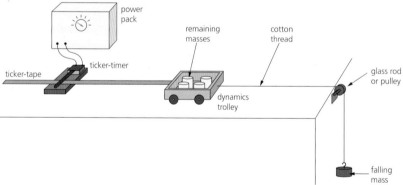

Figure 5.30

Look at Figure 5.30. The falling mass is attached to the trolley by a thread. If the trolley moves to the right it will pull the tape through the timer so the speed can be measured.

1. **Describe the motion of the trolley when the mass hanging on the string is released.**

2. **What force makes the trolley move? What is true about the size of this force during this experiment?**

3. **How could the force on the trolley be altered?**

KEY INFORMATION

With both the ticker-timer and light gate apparatus, the acceleration of the trolley is calculated by measuring its speed at two points and the time taken to go between the two points. A data logger can be set up to measure speed directly, from the time it takes a card attached to the trolley to pass through the light gate. With a ticker-timer, the separation in time between dots on the ticker-tape is 1/50th of a second (0.02 s). Measuring the distance between each dot and dividing this by the time (0.02 s) gives the velocity of the trolley at one point.

4 When we are altering the force, what can we do with the remaining masses to make sure the mass of the system doesn't change?

5 Explain how the mass of the trolley could be altered.

Analysing the results

When analysing results we look for patterns or similarities and draw these together in a conclusion. Adam's group are seeing if their data confirmed that the acceleration of the trolley was proportional to the force acting on it. They used the equipment in Figure 5.30, altering the force on the trolley but keeping the mass on the trolley constant. The table shows their results.

6 What would you expect Adam to find happened to the acceleration when he increased the force acting on the trolley and kept the mass of the system constant?

7 Do his group's results show this?

8 Plot a graph of force against acceleration. What shape is it?

9 What does this show?

Amy's group were doing a different investigation. They kept the force acting on the trolley constant but altered the mass of the trolley. Their results are shown in the table below.

Force (N)	Acceleration (m/s^2)
1	0.5
2	0.9
3	1.4
4	2.1
5	2.5
6	3.0
7	3.6
8	4.1

Mass of trolley (g)	Acceleration (m/s^2)
20	5.0
30	3.3
40	2.5
50	2.0
60	1.7
70	1.4
80	1.2

> **These pages are designed to help you think about aspects of the investigation rather than to guide you through it step by step.**

> **REMEMBER!**
> We can show relationships of direct or inverse proportionality both by calculation and by graphing.

10 What is happening to the acceleration of the trolley as its mass is increased?

11 What shape would the graph of mass against acceleration have?

Exploring the connection with accepted theory

Newton's second law of motion says that if a force accelerates an object, the acceleration is **directly proportional** to the force and **inversely proportional** to the mass of the object. We should be able to analyse the results to see if they show this relationship.

12 Explain how Adam's and Amy's results support this.

13 If the force is tripled and the mass is doubled, what happens to the acceleration?

Newton's third law

Learning objectives:

- identify force pairs
- understand and be able to apply Newton's third law.

KEY WORDS

Newton's third law
weight

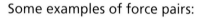

Soldiers firing a large gun never stand behind the gun because the gun moves backwards when it is fired (Figure 5.31).

Force pairs

When two objects, A and B, interact and A exerts a force on B, there is always another force: B exerts a force on A. The forces are equal in size but act in opposite directions and on different objects.

Some examples of force pairs:

- When you walk, your foot pushes the ground backwards; the ground pushes your foot forwards.
- If you stand on a skateboard next to a wall and push on the wall, the wall pushes back on you.
- In a small boat, if you push on the quayside, it pushes back on you and you and the boat moves away.

Figure 5.31 A cannon being fired

1. **What can you say about each pair of forces?**

2. **The tyres of a car push on the road. Name the other force in the force pair.**

Newton's third law

Newton's third law states that whenever two objects interact, the forces they exert on each other are equal and opposite. When two cars collide, the forces on each are of equal size but in opposite directions (Figure 5.32):

$F_1 = F_2$.

Figure 5.32 Forces in a collision

The forces in the pair:

- are the same size
- act in opposite directions
- act on *different* objects.

COMMON MISCONCEPTION

When a pair of objects produce a force on each other we draw the forces as vectors.
But unlike the forces in a free-body diagram, the forces in a Newton's third law pair act on *different* objects.

3 Imagine two people, each sitting on an office chair with wheels. One pushes against the other.

   a Describe what happens next.

   b Explain how this illustrates Newton's third law.

4 Kim and Ben are standing on ice. Kim pushes Ben with a force of 50 N. What force acts on Kim? Describe what happens next.

5 A lorry collides head on with a small car. Compare:

   a the resultant forces acting on the lorry and the car

   b the decelerations of the lorry and the car.

6 An astronaut is floating in space and has become detached from her spaceship. She is wearing a backpack that she is able to remove from her suit. Suggest what she could do to get back to the ship.

Further ideas about force pairs

It is often useful to realise that the pair of forces in Newton's third law are between two objects only. If object A exerts a force on object B then object B exerts a force of equal size but in the opposite direction on object A. The two forces must be the same type of force.

7 A cat is falling to the ground having jumped out of a tree. Ignoring air resistance, the only force acting on the cat is its weight.

   a Complete the sentence by selecting the correct word.

   The force on the cat is due to the Earth pulling on the cat. This type of force is a **magnetic / electrostatic / gravitational** force.

   b Describe the other force in the force pair in this situation.

8 The cat is now sitting on a table.

   a What is the other force in the force pair to the cat's weight?

   b State the other pair of forces which involve the cat.

9 The cat jumps off the table and falls to the ground. Describe how the cat affects the motion of the Earth as it is falling.

10 Describe all of the forces that occur between your body and the Earth when you jump upwards, reach a maximum height and land again.

REMEMBER!

Remember that two vehicles experiencing the same braking force will decelerate by different amounts if they have different masses.

KEY INFORMATION

The normal contact force when two surfaces are in contact is one of a force pair. But Newton's third law also applies for non-contact forces, for example due to gravitational attraction.

Figure 5.33 By Newton's third law, the boy exerts a gravitational force upwards on the Earth of the same size as the gravitational force downwards on the boy (**weight**)

Momentum

KEY WORDS

conservation of
momentum
crumple zones
momentum
rate of change

Learning objectives:

- explain what is meant by momentum
- apply ideas about rate of change of momentum to safety features in cars
- use momentum calculations to predict what happens in a collision.

Understanding momentum enables car designers to save lives.

HIGHER TIER ONLY

Understanding momentum

A moving object has **momentum**. The amount of momentum an object has depends on its mass and its velocity:

momentum = mass × velocity

$p = mv$

where p is the momentum in kg m/s, m is the mass in kg and v is the velocity in m/s.

Example: Evie has a mass of 60 kg. Her momentum is 120 kg m/s. How fast is she moving?

Momentum = mass × velocity

$$\text{velocity} = \frac{\text{momentum}}{\text{mass}} = \frac{120 \text{ kg m/s}}{60 \text{ kg}} = 2 \text{ m/s}$$

KEY INFORMATION

Velocity is a vector so momentum is also a vector.

1 Calculate the momentum of a car with a mass of 1000 kg travelling at 20 m/s.

2 Tom has a mass of 60 kg. His momentum is 240 kg m/s. How fast is he moving?

Changes in momentum

In an accident a car stops suddenly. Its momentum becomes zero. The larger the force applied to the car, the quicker its momentum becomes zero.

We can calculate this by combining ideas we have met already:

force, $F = ma$ (Newton's second law)

$$\text{acceleration} = \frac{\text{change in velocity}}{\text{time}} = \frac{(v - u)}{t}$$

Therefore, $F = \dfrac{m(v - u)}{t}$ or $F = \dfrac{m\Delta v}{\Delta t}$

since change in momentum = $m\Delta v$

MATHS

$m\Delta v$ is a change in momentum (we assume the mass stays the same). $m\Delta v/t$ is the **rate of change** of momentum – how quickly the momentum changes.

Example: A car of mass 1000 kg is moving at 12 m/s when it hits a wall and comes to rest. The force on the car is 4800 N. Calculate the stopping time.

$F = \dfrac{m\Delta v}{\Delta t}$ so $\Delta t = \dfrac{m\Delta v}{F}$

$\Delta v = 12$ m/s $- 0$ m/s $= 12$ m/s, $m = 1000$ kg and $F = 4800$ N

$\Delta t = \dfrac{1000 \text{ kg } (12 \text{ m/s} - 0)}{4800 \text{ N}} = \dfrac{12\,000 \text{ kg m/s}}{4800 \text{ N}} = 2.5$ s

crumple zones   crumple zones

Safety features are designed to increase the time a car or its passengers takes to stop. It means that momentum has not been changed as quickly. The longer the time the smaller the force on the car's occupants.

Figure 5.34 Crumple zones in a car

Crumple zones increase the time between first impact and the car stopping (Figure 5.34). The rate of change of momentum is smaller, reducing the force on the car's occupants.

3 **An airbag inflates on impact. Explain with reference to momentum how this improves safety.**

4 **Suggest why there is a greater risk to the driver and front seat passenger if a back seat passenger fails to wear a seat belt.**

5 **A car of mass 800 kg is moving at 12 m/s when it collides with a wall. The force on the car is 3000 N. Calculate the stopping time.**

Conservation of momentum

We can use the principle of **conservation of momentum** to calculate velocities before and after a collision. The principle states that, in a closed system, the total momentum before a collision is equal to the total momentum after the collision.

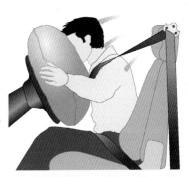

Figure 5.35 A seat belt stretches a little so the passenger's rate of change of momentum is reduced in a crash

In Figure 5.36, the total momentum of the two trolleys before the collision is equal to the total momentum of the two trolleys after the collision, as long as the system is closed.

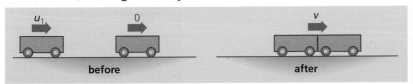

u_1 0 v

before   after

Figure 5.36 An inelastic collision

> **KEY INFORMATION**
>
> A closed system means that no external forces act on the objects involved in the event.

Example: A trolley has a mass of 3 kg and is travelling at 3 m/s. It collides with a second trolley which has a mass of 2 kg and is travelling at 2 m/s. They stick together. Find their common velocity.

$m_1 u_1 + m_2 u_2 = (m_1 + m_2)v$

where u_1 and u_2 = the velocities before the collision and v = common velocity after the collision.

$(3 \text{ kg} \times 3 \text{ m/s}) + (2 \text{ kg} \times 2 \text{ m/s}) = (3 \text{ kg} + 2 \text{ kg})v$

$13 \text{ kg m/s} = 5v \text{ kg}$

$v = \dfrac{13}{5}$ m/s $= 2.6$ m/s to the right

6 **A car of mass 1200 kg travelling at 30 m/s runs into the back of a stationary lorry. The car and lorry move at 4 m/s after impact. Determine the mass of the lorry.**

7 **A truck of mass 2 kg travels at 8 m/s towards a stationary truck of mass 6 kg. After colliding the trucks move off together. Calculate their common velocity.**

Keeping safe on the road

Learning objectives:

- explain the factors that affect stopping distance
- explain the dangers caused by large deceleration
- estimate the forces involved in the deceleration of a road vehicle
- apply the idea of rate of change of momentum to explain safety features.

Road safety campaigns use slogans such as 'Kill your speed not a child!' What difference does reducing speed make in the case of an accident?

Driving safely

A car driver cannot begin to stop a car immediately. It takes time for the driver to react to a stimulus. This is called **reaction time**. The greater the speed of the car the further it travels while the driver thinks (Figure 5.37).

Thinking distance is the distance travelled during the reaction time. It is the distance travelled between the driver seeing a danger and taking action to avoid it, such as braking.

Braking distance is the distance travelled before a car stops after the brakes have been applied. It increases as the speed of the car increases.

Stopping distance = thinking distance + braking distance (Figure 5.38).

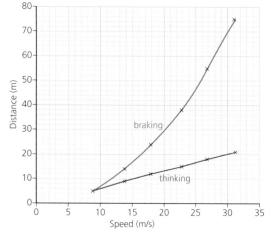

Figure 5.37 How speed affects stopping distance, for a typical car in good conditions

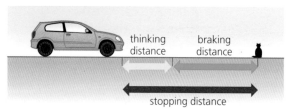

Figure 5.38 Stopping distance

> **REMEMBER!**
>
> Don't confuse thinking distance and braking distance.

1. Laura's thinking distance is 12 m. Use Figure 5.37 to determine:
   a her speed
   b her braking distance
   c her stopping distance.
2. Joe is driving at 80 km/h (22 m/s). Use Figure 5.37 to find his stopping distance in m.

Factors that affect stopping distance

Reaction time varies between 0.2 s and 0.9 s, but is typically about 0.7 s. Reaction time and hence thinking distance may increase if a driver is

- tired
- under the influence of alcohol or other drugs
- distracted or lacks concentration.

> **KEY INFORMATION**
>
> One method of measuring human reaction time is to try catching a ruler as soon as you see it begin to fall. You will need to use the distance the ruler has dropped to calculate the time that has passed.

The braking distance may increase:

- when the road is wet or icy
- the car has poor brakes or bald tyres
- the speed of the car is greater.

Figure 5.37 shows how important it is to keep a suitable distance from the car in front.

3 In Figure 5.37, how does braking distance increase as speed increases?

4 Explain why a driver should slow down when road conditions are poor.

The dangers caused by large deceleration

Kinetic energy is transferred away from a vehicle that is slowing down. Work done by the friction force between the brakes and the rotating wheel reduces the kinetic energy of the vehicle. Energy is transferred to the brake pads, which heat up.

Many journeys involve slowing down to rest from a high speed and do so without discomfort or injury. However, it is when the vehicle needs to stop from a very high speed in a short time (a large deceleration) that problems can arise.

If deceleration is very rapid, it means a large braking force is being applied. This can result in the brake pads overheating or the driver losing control of the direction of the vehicle.

DID YOU KNOW?

Cycle helmets are made from materials that take time to deform, to decrease the rate of change of momentum in an accident.

HIGHER TIER ONLY

When a car decelerates rapidly, this can be dangerous for a driver or passenger. A large deceleration means the momentum of passengers is reduced very quickly. A rapid change in momentum can cause injury due to the large force exerted on the person. The purpose of seat belts and air bags is to bring the passenger to rest over a longer time period than that taken by the car (see topic 5.12).

To estimate the force needed to stop a vehicle, you could either calculate the deceleration and then find the force from $F = ma$, or calculate how much kinetic energy ($\frac{1}{2}mv^2$) must be transferred away from the car and equate this to the work done (Fs) by the braking force.

Example: A lorry is travelling on the motorway. Estimate the braking force needed to stop the lorry in a distance of 100 m.

We need to estimate the following values: mass of lorry (40 000 kg), initial speed of lorry (25 m/s) and braking distance of lorry (100 m).

$v^2 - u^2 = 2as$

$0 - 25^2 = 2 \times a \times 100$

$a = -625/200 = -3.125$ m/s^2

force = mass × acceleration

= 40 000 kg × −3.125 m/s^2

= −125 000 N or −125 kN

5 A car is travelling on the motorway. Use the idea of work done to estimate the braking force needed to stop the car safely. Give your answer to three significant figures.

6 Explain why a driver who repeatedly decelerates rapidly might increase the stopping distance of the car.

7 Children's playgrounds have special ground surfaces that take time to deform. Explain how this feature helps protect children from harm if they fall to the ground.

Moments

Learning objectives:

- describe the turning effect of a force about a pivot
- explain and use the principle of moments
- explain how the turning effect on an object depends on the position of its centre of mass.

KEY WORDS

centre of mass
moment
newton-metre
   (Nm)
pivot

We're used to thinking of forces moving things from one place to another but they can also be used just to turn things round.

Turning effect of a force

It is difficult to tighten a nut onto a bolt with your fingers, especially if the thread is a bit rusty. With a spanner you can use a force to produce a much larger turning effect (Figure 5.39).

The scientific name for a turning effect is a **moment**. The point the moment acts around is called the **pivot**. The size of the moment depends on the size of the force (in newtons) and the perpendicular distance from the pivot to the line of action of the force (in metres).

Figure 5.39 A longer spanner and a bigger force give a larger moment

You can calculate the size of the moment using the equation:

moment = force × perpendicular distance from the pivot to the
line of action of the force

$M = Fd$

The unit of a moment is the **newton-metre (Nm)**. Moments are described as *clockwise* or *anticlockwise*, depending on their direction. A pivot can also be called a fulcrum.

Example: A cyclist pushes down on the pedals with a vertical force of 300 N. The pedal length is 18 cm. What is the moment of the force when the pedal is:

a horizontal   **b** vertical?

a When the pedal is horizontal, its length is perpendicular to the downwards force applied by the cyclist.

Moment = force × perpendicular distance from the pivot to
the line of action of the force

= 300 N × 0.18 m

= 54 N

b When the pedal is vertical, the line of action of the downwards force applied by the cyclist passes through the pivot, so the moment of the force is zero.

> **KEY INFORMATION**
>
> The force is in newtons and the distance is in metres, therefore moments are measured in newton-metres (Nm).

1. **a** Calculate the size of the moment shown in **Figure 5.40**.

   b What changes could you make to increase the moment of the force in **Figure 5.40**?

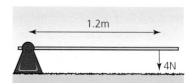

Figure 5.40 The moment acts about the pivot

2 A force of 5 N is applied 1.5 m from a pivot. Calculate the moment of the force about the pivot.

3 Explain, using ideas about moments, why a door handle is placed well away from the hinge.

The principle of moments

When a beam is balanced we say it is in equilibrium. In this case the sum of the clockwise moments about a point (the balance point) is equal to the sum of the anticlockwise moments about that point.

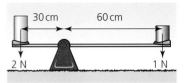

Figure 5.41 Forces on a balanced beam

Example: A load of 200 N is placed on a beam at a distance of 0.2 m from the pivot (Figure 5.42). A person pushes down on the other end of the beam, at a distance of 0.4 m from the pivot. What force, F, is needed to balance the load?

Total clockwise moment about the pivot	=	total anticlockwise moment about the pivot

$$200 \text{ N} \times 0.2 \text{ m} = F \times 0.4 \text{ m}$$

$$F = \frac{40 \text{ Nm}}{0.4 \text{ m}} = 100 \text{ Nm}$$

4 Show that the beam in Figure 5.41 is balanced.

5 Calculate X in Figure 5.42.

6 In Figure 5.43, $F_1 = 6$ N, $d_1 = 20$ cm and $d_2 = 40$ cm. The beam is balanced. Calculate F_2.

7 In Figure 5.43 F_1, d_1 and d_2 remain the same values as in question 6. However, a 10 N force that acts upwards is added 10 cm to the left of the pivot. Calculate the value of F_2 which balances the beam.

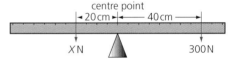

Figure 5.42

Figure 5.43

Centre of mass

Everything is made up of lots of tiny particles, each with a small gravitational force on it. A metre rule balances when suspended at one particular point. If the metre rule is uniform this will be the centre of it. This is because the gravitational forces have turning effects about this point which cancel out.

The weight of the object can be considered to act at a single point referred to as the object's **centre of mass**.

8 Look at the ruler in Figure 5.42. Explain where the centre of mass is.

An object becomes unstable when the vertical line through its centre of mass falls outside its base, which acts as a pivot. The weight of the object causes a turning effect about this pivot.

9 Why does vase C topple in Figure 5.44?

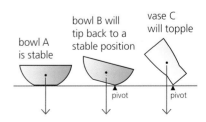

Figure 5.44 The weight of each bowl or vase is shown as a force vector passing through the object's centre of mass

Levers and gears

Learning objectives:

- describe how levers and gears can be used to transmit the rotational effect of a force
- explain how levers and gears transmit forces.

KEY WORDS

gear
lever

When you change down a gear on a bicycle you have to pedal faster to keep moving at the same speed. But the turning effect (or torque) on the rear wheel is increased so it is easier to cycle uphill.

Levers

A **lever** uses the moment of a force. In Figure 5.45 a force pushing down at one end (the effort force) gives a clockwise moment about the pivot. This is greater than the anticlockwise moment caused by the load at the other end, so the load is lifted. In this example the load is much bigger than the effort force. However, the load only moves a small distance compared to the effort force.

DID YOU KNOW?

Stonehenge was built 4000 years ago using huge rocks from over 160 km away. How do you think the rocks were moved?

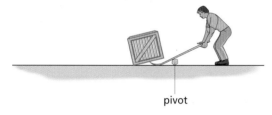

pivot

Figure 5.45 A large load can be moved by a smaller effort

KEY INFORMATION

Some machines multiply forces but the distance moved is much smaller. Energy is conserved: you cannot get more out than you put in.

1 **Explain how a bottle opener acts as a lever.**

2 **Explain how a wheelbarrow (Figure 5.46) makes lifting a load easier.**

Levers let us move a large force using a small one but the distance moved is much smaller. Energy transferred to the load is the increase in the gravitational potential energy:

$E_p = mgh$

= weight of load × vertical height increase

The work done by the effort force is:

$W = Fs$

= effort force × distance moved along the line of action of the force

By the principle of conservation of energy these must be equal, so if the effort force is smaller than the load force, the distance moved by the effort force must be greater than the distance moved by the load.

effort

load

fulcrum

Figure 5.46 Wheelbarrow

Gears

Gears are used for transmitting (passing on) the rotational effect of a force from one part of a machine to another. In a bicycle, for example, gear wheels (with the help of a chain) transfer a force from the pedals to the back wheel (Figure 5.47). Both gear wheels turn in the same direction.

The force (effort) applied to the pedals rotates the first gear wheel. If both gear wheels are the same size, they rotate at the same speed.

When a cyclist starts to climb a steep hill they change gear. This moves the chain to a larger gear wheel on the back wheel. The larger gear wheel always moves more slowly but with a larger force. The cyclist pedals quickly but moves more slowly.

If the gear wheels are the other way around, the smaller gear wheel (driving the back wheel) moves faster but with a smaller force. This is the arrangement a cyclist uses to cycle downhill.

Figure 5.47 Gears transfer a force from the pedals to the back wheel

3 How do gears help a cyclist to climb a steep hill?

4 Why would the cyclist not want to stay in the same gear to come downhill?

Linked gear wheels

When two gears are in contact, the direction of the rotation can be changed. In Figure 5.48, when the smaller wheel moves clockwise the larger wheel moves anticlockwise. Again, the smaller gear will rotate faster than the larger gear. The larger gear rotates more slowly but with more force.

5 An 'idler' gear has two linked gears of the same size. Why might this be useful?

DID YOU KNOW?

In a car, gears transmit power from the crankshaft (the rotating axle) that takes power from the engine to the driveshaft running under the car, which ultimately powers the wheels.

Figure 5.48 Linked gear wheels

Pressure in a fluid

Learning objectives:

- explain how pressure acts in a fluid
- calculate pressure at different depths in a liquid
- explain what causes upthrust.

Submarines can only dive safely to a certain depth. If they go deeper than that, the water pressure is too great for the hull to withstand and it gets crushed.

Pressure in fluids

The **pressure** in fluids causes a force normal to (at right angles to) any surface. This means the pressure in a fluid acts in all directions (Figure 5.49). A fluid can be either a liquid or a gas.

To calculate pressure we need to work out how much force is being applied over a certain area. We use an area such a square metre, or a square millimetre. We refer to this as the unit area.

Pressure is the force acting normal to a unit area of a surface.

$$\text{Pressure, } p = \frac{\text{force normal to a surface}}{\text{area of that surface}}$$

$$p = \frac{F}{A}$$

Pressure is measured in **pascals** (Pa) when the force is in N and the area is in m^2. 1 Pa = 1 N/m^2. A pascal is a small unit so we often use kilopascals (kPa).

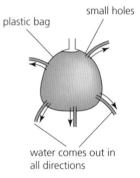

Figure 5.49 Pressure in a fluid acts in all directions

HIGHER TIER ONLY

Pressure at different depths in a liquid

The deeper you go in water, the greater the pressure becomes. This is because the deeper you are, the greater the weight of the water above you. If you dive into deep water, your ears may hurt because the pressure outside your ears is much greater than the pressure inside your ears.

1. If you make holes at different heights on the side of a tall container and fill it with water (Figure 5.50), why does water come out further from the lower holes?

2. Imagine taking a sealed tin can on a dive. As you take it down, deeper and deeper, think about the pressure acting on it. Explain what will happen to the can if you take it deep enough.

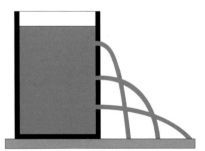

Figure 5.50 Pressure increases with depth

The pressure at a certain point in a liquid depends on the height of liquid above the point (h), the **density** of the liquid (ρ) and the gravitational field strength (g).

Pressure due to a column of liquid

= height of column × density of liquid

× gravitational field strength, g

$p = h\rho g$

3 Calculate the pressure due to the petrol 5 m below the surface of a tank of petrol, using values from the table. Take g as 9.8 m/s².

4 A rectangular tank is filled with water to a depth of 3 m. Calculate the pressure at the bottom of the tank due to the water. Take g as 9.8 m/s².

> **KEY INFORMATION**
>
> To calculate the *difference* in pressure at different depths in a liquid, such as the pressure difference between the upper and lower surfaces of a submerged object (Figure 5.51), use $\Delta p = \rho g \Delta h$: Δh is the vertical height difference.

Floating and sinking

A submerged object experiences a greater pressure on the bottom surface than on the top surface. This creates a resultant force upwards. This force is called the **upthrust**.

When solid objects are placed in a liquid they will float or sink depending on the density of the solid material compared with the density of the liquid. If the object is denser than the liquid it will sink because the weight is greater than the upthrust. If the object is less dense than the liquid the upthrust is greater than the object's weight and it will float to the surface. If the object has exactly the same density as the liquid it will stay still, neither sinking nor floating upwards. The weight is equal to the upthrust. For example, ice floats on water because it is less dense than water. Lead sinks in water because it is denser than water.

Substance	Density (kg/m³)
lead	11 000
water	1000
ice	920
air	1.3
petrol	800

5 Use the values in the table to name a substance that sinks in water.

6 Use the values in the table to name a substance that floats in water.

7 Would ice float on petrol? Explain your answer.

8 How are the densities in the table related to the states of matter?

9 A 1 m³ solid cube is held upright beneath some water. The base of the cube is 10 m below the water's surface. (g = 9.8 m/s²)

a Calculate the pressure due to the water at the base of the cube.

b Calculate the pressure due to the water at the top of the cube.

c Explain why there is an upthrust.

d The weight of the cube is 12 000 N. Determine whether the cube floats or sinks.

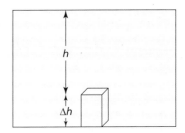

Figure 5.51 Pressure in a liquid

Atmospheric pressure

Learning objectives:

- show that the atmosphere exerts a pressure
- explain variations in atmospheric pressure with height
- describe a simple model of the Earth's atmosphere and atmospheric pressure.

The atmospheric pressure is about 100 000 N/m^2 at sea level but the higher you go, the smaller it becomes. The air becomes thinner and thinner and space begins. Space begins about 80 km from the surface of the Earth.

Atmospheric pressure

The air exerts a pressure. This causes a force normal (at right angles) to any surface. We can show how strong this force is by pumping the air out of a metal can (Figure 5.52). To start with, there is air on the inside pushing out and air on the outside pushing in. Removing the air from the inside means it is only the strength of the can itself that is resisting the pressure on the outside, and it soon collapses.

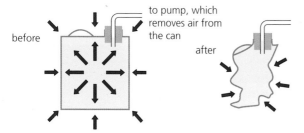

before

to pump, which removes air from the can

after

Figure 5.52 **Atmospheric pressure** can crush objects

1 **a** Explain what happens when you suck the air out of a paper bag in terms of pressure.

b Why is it wrong to say 'I've sucked the sides in'?

2 Imagine having a balloon inflated with helium.

a What is applying pressure from the inside?

b What is applying pressure on the outside?

DID YOU KNOW?

....

Atmospheric pressure is not constant but varies slightly. These changes are a very useful predictor of weather.

Living at the bottom of a 'sea' of air

In the same way that pressure varies with depth in liquids, so too does the pressure in the atmosphere. The atmospheric pressure is caused by the amount of air above. If you go up a mountain the atmospheric pressure decreases. This is because you are going nearer to the top of our 'sea' of air. The air pressure is less.

Many people live close to sea level, at the bottom of the atmosphere.

3 Why does the atmospheric pressure decrease if you go up a mountain?

4 Suggest why most of the air in our atmosphere is close to the Earth's surface.

5 Explain why the helium balloon in question 2 expands when it is released. (The balloon is less dense than the air.)

Explaining what causes atmospheric pressure

In a gas the particles are continuously moving around at high speed. There are many collisions. Sometimes the particles bump into each other and sometimes they hit the sides of the container they are in (Figure 5.53). It is these collisions with a surface that cause pressure. It is the collision of molecules in air with a surface that creates atmospheric pressure.

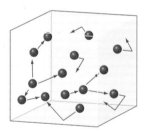

Figure 5.53 A gas exerts a pressure on the wall of the container

If the density of air is greater, there will be more weight of air above a surface. This means there will be more particles in a given space and so more collisions per second with a surface; the pressure is greater. Higher up in the atmosphere the air is less dense and there are fewer particles in a given volume. This means fewer collisions per second and less pressure on a surface.

6 How does a gas exert a pressure?

7 Apply the particle model of gas pressure to explain why an evacuated can collapses.

8 A man has climbed a skyscraper just using suction cups to hold him to the wall. Suction cups are airtight so no air can move in to fill the gap between them and the wall. Explain, in terms of the particle model, why the man could stick to the wall.

> **KEY INFORMATION**
>
> Collisions between air molecules do **not** contribute to the atmospheric pressure, only collisions of molecules in air with a surface, such as your skin.

Forces and energy in springs

Learning objectives:

- explain why you need two forces to stretch a spring
- describe the difference between elastic and inelastic deformation
- calculate extension, compression and elastic potential energy.

KEY WORDS

compression
elastic deformation
elastic potential energy
extension
inelastic deformation
limit of proportionality
linear
non-linear
spring constant

Some skyscrapers are mounted on springs. This prevents the building from shaking too much in an earthquake. The builders need to make sure that the springs are strong enough to withstand the weight of the building.

Forces on a spring

The spring in Figure 5.54 is being stretched by two forces. The weight of the mass is pulling the spring downwards. However, if this was the only force acting on the spring then it would accelerate towards the ground. The spring is in equilibrium. The other force is from the bar, which is pulling the top of the spring upwards.

1. What is the resultant force on the spring if it is at rest?

2. Determine the force at the top of the spring when the weight of the mass is 3 N.

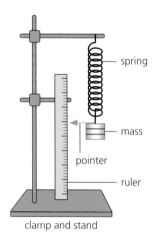

Figure 5.54 Forces stretching a spring

Forces can stretch and squash a spring elastically or inelastically. **Elastic deformation** occurs when the spring returns back to its original length when the forces are removed. A spring that is permanently altered has undergone **inelastic deformation**.

3. What would happen to a bed mattress if its springs had been inelastically deformed?

The **extension** or **compression** of the spring is how much its length changes when the forces are applied. You measure extension when the spring is being stretched and compression when it is being squashed.

4. A spring has an original length of 6.0 cm. It is then squashed to a length of 4.8 cm. Determine the compression of the spring.

The relationship between force and extension

You can see that the graph in Figure 5.55 is a straight line up to a point which is called the **limit of proportionality**. This

DID YOU KNOW?

Scientists have developed a material that can elastically deform to 20 times its original length without breaking.

shows that there is a **linear** relationship between the force and extension when the force applied is small.

Since the straight line goes through the origin, we can also say that force is directly proportional to the extension.

You can calculate the force or the extension using the equation:

force = **spring constant** × extension

$F = ke$

where k, the spring constant, is measured in N/m and depends on the spring; e can be the extension or the compression and is measured in m.

Since $F = ke$, $k = F/e$ and the spring constant for an elastic material can be found from the gradient of the straight part of a force–extension graph. Note, however, that the graph in Figure 5.55 is plotted for extension against force, so the gradient of this graph is $1/k$.

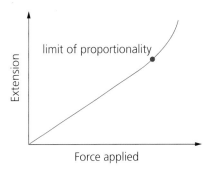

Figure 5.55 The relationship between force and extension for an elastic object, such as a spring

5. A spring has an extension of 0.08 m when it is stretched by a force of 4 N. What is its spring constant?

6. A spring has a spring constant of 2000 N/m and is 10 cm long. What force is needed to compress the spring to a length of 4 cm?

Beyond the limit of proportionality, the line in Figure 5.56 begins to curve. The relationship between force and extension is now **non-linear**.

Elastic potential energy

Doing work on the spring transfers energy into the **elastic potential energy** store of the spring. The amount of work done equals the amount of elastic potential energy stored. This energy can be transferred to other energy stores when the spring returns to its original length.

If the limit of proportionality is not exceeded then you can calculate the elastic potential energy from the equation:

elastic potential energy
= 0.5 × spring constant × (extension or compression)²

$$E_e = \frac{1}{2} ke^2$$

Example: A block is attached to a spring. The spring constant is 28 N/m and the total mass of the block and spring is 50 g. What is the maximum vertical height the block can reach after the spring is compressed by 5 cm then released? ($g = 9.8$ N/kg)

$$E_e = \frac{1}{2} ke^2 = \frac{1}{2} \times 28 \times (0.05)^2 = 0.035 \text{ J}$$

Provided no energy is dissipated, the decrease in the elastic potential energy store equals the increase in the gravitational potential energy store:

$$E_p = mgh = 0.035 \text{ J}$$

$$h = \frac{0.035}{(0.05 \times 9.8)} = 0.07 \text{ m}$$

KEY INFORMATION

Formulae involving the **spring constant** only work if the limit of proportionality is not exceeded.

7. Calculate the elastic potential energy stored by a spring that has been stretched by 2 cm. The spring constant is 30 N/m.

8. A spring has a spring constant of 10 N/m. Determine its extension once 0.2 J of work is done stretching it. Assume the limit of proportionality is not exceeded.

9. A 10.0 g spring is stretched by 10.0 cm. It is then released and flies 1.0 m upwards before it starts to fall down again. Assuming all of the elastic potential energy transfers to gravitational potential energy, estimate the spring constant of the spring. What else have you assumed in your calculation? ($g = 9.8$ N/kg)

REQUIRED PRACTICAL

Investigate the relationship between force and the extension of a spring

Learning objectives:

- interpret readings to show patterns and trends
- interpret graphs to form conclusions
- apply the equation for a straight line to the graph.

A class of students have been making and testing springs. They have then investigated how these springs behave when loaded with a weight. They have carried out the experiment and gathered the data. Their next job is to see what it shows.

These pages are designed to help you think about aspects of the investigation rather than to guide you through it step by step.

Analysing the results

When analysing results look for patterns or similarities and see if these can be used to draw a conclusion.

Ellie investigated three springs and recorded her data in a table. The table shows the readings and the calculations.

Force (N)	Spring 1 extension (mm)				Spring 2 extension (mm)				Spring 3 extension (mm)			
	1st reading	2nd reading	3rd reading	Mean	1st reading	2nd reading	3rd reading	Mean	1st reading	2nd reading	3rd reading	Mean
10	8	9	7	8	3	3	4	3	4	4	4	4
20	20	13	12	15	6	7	6	6	8	9	10	9
30	20	19	30	23	10	10	11	10	14	13	12	13
40	34	30	32	32	13	13	13	13	16	16	16	16
50	35	45	40	40	16	17	17	17	22	22	20	21

1. What apparatus should be used to measure the force applied to the spring?

2. How should the **extension** of the spring be measured?

3. How did Ellie calculate the **mean** values?

4. Are there any **anomalies**?

5. Suggest whether these calculations are a good way of getting close to the true values of the extension.

DID YOU KNOW?

The relationship between the force on a spring and the extension it produces was first put forward by Robert Hooke in 1660. He was an inventor, a scientist and an architect. He invented the universal joint now used in the transmission systems of motor cars, the iris diaphragm used in cameras and he was first to coin the word 'cell' in biology. He designed many of the new buildings after the Great Fire of London.

Making sense of the graph

Darren's group have drawn a graph showing extension against the weight hung from the spring for one of their springs (Figure 5.56). They are now looking at the graph to see what conclusion they can draw.

6 What does the straight part of the graph show?

7 What happened in the curved part of the graph?

8 What conclusion could Darren draw from this graph?

Using the equation of a straight line on a graph

Katie's group have plotted the graphs from the results of their experiments for three springs on a graph (Figure 5.57). They know that any straight line graph can be described using the general equation $y = mx + c$ and they are applying it to these lines.

With the first of these (A) they can see that:

- x is the force, which changes from 0 N to 8 N
- y is the extension, which changes from 0 cm to 16 cm
- c is the y-intercept, which is 0 as the line crosses the y-axis at (0,0).

Therefore $16 = (m \times 8) + 0$, so the gradient, $m = 2$.

Since the intercept on Figure 5.57 is zero, the extension is directly proportional to the force. Since:

force applied to a spring = **spring constant** × extension

extension = force applied to a spring/spring constant

so the gradient of the graph is 1/spring constant.

The spring constant can be found from 1/gradient, or 1/m.

9 Calculate the value of m for springs B and C.

10 Use the equation $k = F/e$ to calculate the spring constant for springs B and C. Explain which of the springs was the stiffest.

11 The spring constant can also be found by redrawing the graph so that extension is on the horizontal axis and force on the vertical axis. The spring constant is then the gradient of the line for each of the springs.

Try doing this and then check your answers against those for the previous question.

12 **a** Determine the area under line A in Figure 5.57. (Make sure the extension is in metres.)

b Calculate the elastic potential energy stored in spring A when a force of 8 N is applied. $E_e = \frac{1}{2}ke^2$.

c Compare your answers to parts (a) and (b) and explain why this is the case.

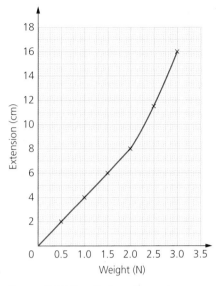

Figure 5.56 Results from Darren's group

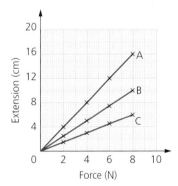

Figure 5.57 Results from Katie's group

REMEMBER!

The spring constant is always measured in N/m so it is important to make sure extension is measured in metres.

KEY CONCEPT

Forces and acceleration

Learning objectives:

- recognise examples of balanced and unbalanced forces
- apply ideas about speed and acceleration to explain sensations of movement
- apply ideas about inertia and circular motion to explain braking and cornering.

KEY WORDS

acceleration
balanced
force
inertia
speed
velocity

The reason why scientists study forces is that they tell us about how things move. If you know all the forces acting on something you can predict very accurately if and how it will move. Newton's laws of motion are very simple but the situations you can apply them to can be complex. The laws were first suggested by Isaac Newton (Figure 5.58) and published in 1686. NASA used these simple laws to get men safely to the Moon and back, and theme parks use them to thrill us on their rides.

Figure 5.58 Isaac Newton

Balanced and unbalanced forces

People who design rides for theme parks understand what will make a ride exciting. It's not just **speed**. Stealth, at Thorpe Park, is the UK's fastest roller coaster and reaches 128 km/h, which is not much faster than a car on a motorway, but is much more exciting. You need high **acceleration** for a good ride and that requires **force**, or, to be precise, unbalanced force.

Newton explained that if all the forces on an object balance out, its speed will be constant and its direction will not change, or it will not move if it is at rest. If the forces are not **balanced** the object's speed or direction will change.

It is the acceleration that makes roller coasters (Figure 5.59) such fun. Newton's first law of motion states that a moving object tends to stay moving. This resistance to change is called **inertia**. When the roller coaster speeds up, the back of the seat pushes you forward, accelerating you. When the roller coaster slows down, your body tries to keep going at its original speed. The harness in front of you pushes your body backward. These pushes and pulls on your body give you the thrills you enjoy.

Figure 5.59 A roller coaster

1. What happens to the acceleration of the roller coaster if the force propelling it forwards increases?

2. Explain what would happen if the roller coaster had a negative acceleration and the passengers weren't wearing restraints.

Figure 5.60 Going around a loop

Looping the loop

Curves and loops on the ride (Figure 5.60) cause the direction you are travelling in to change, so your **velocity** changes. Remember that a change in velocity is acceleration even if the speed remains constant. To change direction, there needs to be a force towards the centre of the circle that the curve is part of. This is an example of Newton's second law.

3 You feel the seat pushing you as you go round the loop. Suggest what would happen if there was no seat.

4 Why do we say that a train going round a loop is accelerating, even when its speed is constant?

5 Use your answer to question 4 to explain the difference between speed and velocity.

6 Newton said that unless there is an unbalanced force, an object will continue in a straight line at a steady speed. Describe how this force is applied to the riders in the roller coaster.

Drop towers

Zumanjaro: Drop of Doom in America is the world's tallest drop tower at 125 m. After a brief pause at the top, you are released and fall under the pull of gravity. You are in free-fall for over 10 s, reaching speeds of over 140 km/h. The force is unbalanced and causes acceleration; the same thing (but much more gently) happens when you descend in a lift.

Figure 5.61 A drop tower

To slow the ride down and stop it, another unbalanced force is needed. This is provided by giant permanent magnets which provide an upward force.

7 If a free-fall ride accelerates at 10 m/s^2 and the loaded car has a mass of 2500 kg, what force is acting?

8 If the braking force at the end of the ride was half the size of the force of gravitational attraction, what acceleration would the riders experience?

9 A passenger on Zumanjaro was holding an open bottle of water. As the ride accelerated downwards he decided to tip the bottle upside down to make the water fall out. Explain what happened to the water.

DID YOU KNOW?

g-force is a way of comparing forces by measuring the acceleration they produce. A force of 1*g* causes acceleration the same as that of gravity (around 9.8 m/s^2). Humans can tolerate greater *g*-forces horizontally than vertically; accelerating downwards rapidly forces blood into the brain and eyes; any more than 2*g* is dangerous.

MATHS SKILLS

Making estimates of calculations

Learning objectives:

- estimate the results of simple calculations
- round numbers to make an estimate
- calculate order of magnitude.

In 1945 the nuclear physicist Enrico Fermi estimated the energy released by the first atomic bomb test by dropping small pieces of paper from his hand. He based his estimate only on the displacement of the scraps of paper (about 2.5 m) while the blast wave was passing and his distance from the site of the explosion (about 15 km).

Making sensible estimates

It is often useful to **estimate** a quantity. To make an estimate, make a sensible approximation. The symbol ~ means 'approximately equal to'.

Example: Estimate the average acceleration of a typical family car that accelerates from rest to a velocity of 100 km/h (about 30 m/s).

Make a sensible approximation of the time taken to accelerate around 6 s:

The initial velocity is zero.

$$\text{Acceleration} = \frac{\text{change in velocity}}{\text{time taken}}$$
$$\sim \frac{(30\,\text{m/s} - 0\,\text{m/s})}{6\,\text{s}}$$
$$\sim 5\,\text{m/s}^2$$

Example: Estimate the sprinting speed for a typical adult.

Make a sensible approximation of the time he or she takes to sprint 100 m: around 15 s.

$$\text{Speed} = \frac{\text{distance}}{\text{time}}$$
$$= \frac{(100\,\text{m})}{15\,\text{s}}$$
$$= 6.67\,\text{m/s}$$

Write estimates to one significant figure.

speed ~ 7 m/s

1 **Estimate the volume of this book. Show your working and assumptions.**

2 **Estimate the running speed for a typical adult over a distance of 500 m. Show your working and assumptions.**

Estimating the result of a calculation by rounding

It is often useful to estimate the result of a calculation without using a calculator. Do this by **rounding** quantities up or down to one significant figure.

Example: Without using a calculator, estimate the deceleration of a van that makes an emergency stop from a speed of 33 m/s. The time taken to brake to a stop is 13 seconds.

$$a = \frac{(v - u)}{t}$$

Substituting rounded values:

$$a \sim \frac{(0 - 30)}{10}$$
$$\sim -3\,\text{m/s}^2$$

3 Without using a calculator, estimate the deceleration of a car that makes an emergency stop from a speed of 21 m/s. The time taken to brake to a stop is 3.5 seconds.

4 For the car whose deceleration you estimated in question 3, use the equation $v^2 = u^2 + 2as$ to estimate the car's braking distance. Then calculate the overall stopping distance, using an estimate for a typical value for the reaction time.

5 Estimate the braking force required to stop the car in question 3. Make a sensible approximation of the mass of the car.

Orders of magnitude

In question 1, two students estimated the volume of this book. Their estimates were 400 cm^3 and 600 cm^3.

Both estimates are of the same order of magnitude. An **order of magnitude** is a factor of 10. When two numbers are the same order of magnitude, this means that one of the numbers is less than 10 times larger than the other.

Example: Make an order of magnitude estimate for the time taken by light to travel across our galaxy, the Milky Way. The diameter of the Milky Way is about 7×10^{17} km to 9×10^{17} km. The speed of light is 3.0×10^8 m/s.

$$\text{Time} = \frac{\text{distance}}{\text{speed}}$$

Using the nearest order of magnitude:

$$\text{time} = \frac{10^{17} \times 10^3 \text{ m}}{10^8 \text{ m/s}}$$
$$= \frac{10^{20}}{10^8}\,\text{s}$$
$$= 10^{12}\,\text{s}$$

6 Make an order of magnitude estimate for the mass of air in this room. The density of air at ordinary atmospheric pressure and room temperature is 1.2 kg/m^3.

7 Make an order of magnitude estimate for the time taken by light to travel across the length of the UK.

KEY INFORMATION

Using rounded values to make an estimate allows you to check that the result you find using your calculator is correct. If your estimate and your calculated answer are not similar, then you may have made a mistake in your calculation, for example when typing in the numbers.

KEY INFORMATION

Use an order of magnitude when the precise value of a quantity is unknown. Make an estimate rounded to the nearest power of 10 and use that in your calculation.

Check your progress

You should be able to:

Know that forces are vectors and have magnitude and direction →	Explain the difference between contact and non-contact forces →	Represent vector quantities by arrows
Understand that average speed = distance/time →	Know that acceleration is the rate at which speed changes →	Calculate acceleration from change in velocity/time taken
Explain the significance of the gradient of a distance–time graph →	Interpret a journey represented on a distance–time graph →	Determine the instantaneous speed from the tangent to a distance–time graph of an accelerating object
Explain the significance of the gradient of a velocity–time graph →	Interpret a journey represented on a velocity–time graph →	Determine total distance travelled from a velocity–time graph
Recall the equation for uniform motion →	Apply the equation for uniform motion →	Rearrange the equation for uniform motion
Draw a free-body diagram to represent forces acting on an object →	Calculate the resultant force acting on an object →	Determine the components of a force
Apply Newton's first law to a stationary object and an object moving in a straight line at a constant speed →	Link Newton's first law to the idea of a zero resultant force →	Explain what is meant by inertia
State Newton's second law and recall the equation $F = ma$ →	Use $F = ma$ to determine force, mass or acceleration →	Explain what is meant by inertial mass
Recognise that weight and mass are not the same →	Explain the difference between weight and mass →	Relate the ideas of weight and mass to Newton's second law
State Newton's third law →	Apply Newton's third law to simple equilibrium situations →	Explain how Newton's third law applies
State that vehicle speed and reaction time affect the stopping distance of a vehicle →	Describe factors that affect a driver's reaction time and a vehicle's braking distance Explain why the temperature of a vehicle's brakes increases during braking →	Interpret a graph that relates speed to stopping distance for different vehicles Calculate braking forces using ideas of stopping distance and energy transfer
Identify measures to increase road safety Explain what is meant by momentum →	Relate measures to increase road safety to ideas about forces and kinetic energy, and to rate of change of momentum →	Explain vehicle safety features in terms of the rate at which momentum is reduced Apply the principle of conservation of momentum to collisions
Explain that a moment is the turning effect of a force →	Calculate the size and direction of a moment Explain how gears and levers transmit the rotational effect of a force →	Apply the idea of moments to contexts such as the balancing of a seesaw
Describe how a fluid exerts a pressure on a surface Describe how pressure varies with depth in a fluid →	Calculate pressure at any depth in a fluid and explain what causes atmospheric pressure →	Explain how a partially (or totally) submerged object experiences upthrust and why atmospheric pressure decreases with height Describe the factors which influence floating and sinking

Worked example

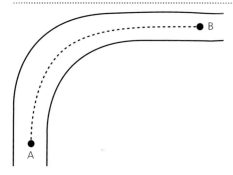

The diagram shows a car travelling round a bend at a constant speed. It took the car 4.5 s to move 90 m from point A to B.

1 **Explain why the distance travelled by the car from A to B is greater than the car's displacement at B.**

The car's displacement is the distance from A to B in a straight line. The car travelled a further distance as it was going round the curve.

This is correct. You should also note that distance is a scalar quantity but displacement is a vector quantity with size and direction.

2 **Calculate the distance the car travelled from A to B.**

$s = vt$ so $v = st$

$v = 90 \times 4.5$

$= 405 \, m/s$

The student correctly remembered the equation $s = vt$ but failed to rearrange it correctly as $v = \dfrac{s}{t}$ to give the answer 20 m/s.

3 **Explain why the car is accelerating as it travels round the bend at a constant speed.**

Acceleration is a change in velocity and the car's velocity is changing as it goes round the bend.

Velocity is speed in a given direction. As the car goes round the corner, its speed stays the same but the direction is constantly changing, so its velocity is constantly changing which is acceleration.

4 **Explain what is meant by stopping distance for a car and what factors might affect this.**

Stopping distance is the distance the car takes while braking and the distance travelled while thinking to apply the brakes. Ice on the road will cause the brakes not to work and increase the stopping distance.

You should aim to give at least two factors that affect thinking distance and braking distance. Thinking distance can be affected by tiredness, drugs and alcohol. Braking distance can be affected by adverse road conditions such as rain or ice, and worn tyres or brakes.

End of chapter questions

Getting started

1. In what units are forces measured? `1 Mark`

2. Which one of these is not a force? `1 Mark`

   a weight   b air resistance   c mass   d upthrust

3. Give an example of a contact force and a non-contact force. `2 Marks`

4. What unit does the symbol Pa represent and what is it a measure of? `1 Mark`

5. Explain why a force is a vector quantity. `1 Mark`

6. Use the equation $s = vt$ to calculate the average speed of a car when it travels 100 m in 5 s. `1 Mark`

7. On Earth, what is the weight of a 50 kg boy? ($g = 9.8\,N/kg$) `1 Mark`

8. Describe the motion represented by the graph below. `2 Marks`

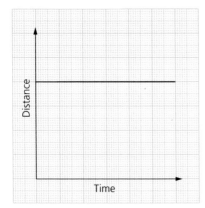

Going further

9 When a see-saw is balanced, what do you know about the clockwise moment? `1 Mark`

10 On a distance–time graph, what is represented by a straight line sloping upwards? `1 Mark`

11 Describe **two** factors that affect the stopping distance of a car. `2 Marks`

12 A force of 80 N is applied to the end of a spanner of length 0.2 m to turn a nut.

   a Calculate the maximum moment the force produces about the centre of the nut. `2 Marks`

   b Explain why a longer spanner would make it easier to turn to nut. `2 Marks`

13 The diagram below shows a can of water with holes in the side. After a short time, the jets of water do not travel as far. Explain why this happens. `2 Marks`

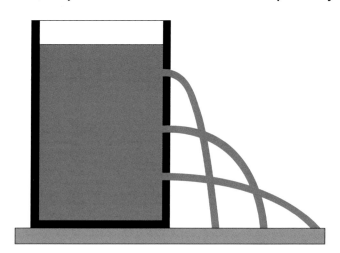

More challenging

14 An object moves in a circle at 2 m/s.

Choose the correct words.   `1 Mark`

The speed of the object:

changes/remains constant

The velocity of the object:

changes/remains constant

15 A car is travelling at a constant speed of 20 m/s. Its mass is 1000 kg and the force of friction opposing the motion of the car is 1500 N.

Determine the size of the force that the engine is providing to drive the car.   `1 Mark`

16 Explain the meaning of limit of proportionality for a spring.   `2 Marks`

17 a Three girls are trying to balance a see-saw which is pivoted in the middle. Ann sits at the end which is 2 m from the middle on one side and Beth sits 1.5 m from the middle on the other side. If Cindy weighs the same as Ann, where on the see-saw should she sit to balance it?   `3 Marks`

b The three girls now stand on a wobbly platform, which is on a spring. The spring constant of the spring is 8000 N. The weight of the platform is 100 N.

Calculate the compression of the spring and explain what you have assumed in the calculation.   `3 Marks`

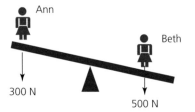

Ann

300 N

Beth

500 N

Cindy

Most demanding

18 What is Newton's third law? `2 Marks`

19 A car of mass 1500 kg is travelling along a motorway at 25 m/s. Calculate the braking distance of the car if the brakes exert a force of 3000 N. `2 Marks`

20 A snooker player hit a white ball with her cue. The white ball then collided with a stationary red ball. The player noticed that the white ball stopped after the collision. She also noticed that the red ball moved with the same velocity as the white ball had before the collision.

Explain the physics behind these observations and make a conclusion about the mass of the balls. Use sensible numerical values to illustrate your answer. Assume no friction acts during the collision. `6 Marks`

`Total: 40 Marks`

WAVES

LIGHT WAVES AND WATER WAVES HAVE SOME THINGS IN COMMON

- All waves carry energy from one place to another.
- When waves hit an object they may be absorbed by it, transmitted or reflected back.
- Waves may change direction (refract) at the point where two different materials meet.

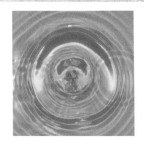

DESCRIBING WAVES

- Frequencies of waves are measured in hertz (Hz).
- Waves travel at different speeds in different materials.
- Sound waves are longitudinal.
- Ripples on the surface of water are transverse.

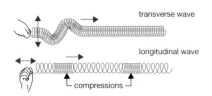

transverse wave

longitudinal wave

compressions

VISIBLE LIGHT

- Sunlight (white light) is made up of a mixture of many different colours.
- Each colour of light has a different frequency.
- Different colours are absorbed or reflected by different surfaces.
- Light waves can travel through a vacuum.

WAVES CAN TRAVEL THROUGH SOLIDS

- Sound waves are produced by vibrations.
- Sound needs a medium to travel through.
- Ultrasound is defined as sound with a pitch too high for humans to hear.

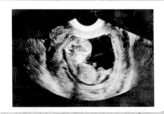

LENSES IN THE EYE AND THE CAMERA

- Light travels in straight lines unless it is reflected or refracted.
- A convex lens refracts light to form an image.
- Your eye contains a convex lens which focuses images on your retina.
- A camera can focus an image on a light-sensitive material such as photographic film.

IN THIS CHAPTER YOU WILL FIND OUT ABOUT:

IN WHAT WAYS DO OTHER ELECTROMAGNETIC WAVES BEHAVE LIKE LIGHT?

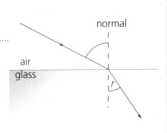

- All electromagnetic waves are transverse waves that transfer energy.
- All electromagnetic waves can be reflected and refracted at a boundary between two different media.

WHAT CHARACTERISTICS OF WAVES CAN BE MEASURED?

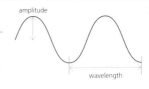

- We can measure the speed, wavelength and frequency of waves.
- We can calculate one of these three properties using the other two.
- The amplitude of a wave is its maximum displacement from its rest position.

ARE THERE ANY WAVES BEYOND THE VISIBLE SPECTRUM?

- The colour of an object depends on which wavelengths it reflects most strongly.
- The visible spectrum is only a small part of a much wider spectrum called the electromagnetic spectrum.
- The invisible waves beyond red are called infrared and those beyond violet are called ultraviolet.
- Gamma rays, X-rays and ultraviolet rays have the highest frequencies (smallest wavelengths) and transfer the most energy.

HOW DO WAVES ALLOW US TO DETECT STRUCTURES WE CANNOT SEE?

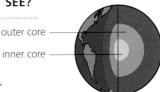

- Echo sounding and ultrasound scans use reflections of high-frequency sound waves to detect objects hidden from view.
- The reflection and refraction arrival times of seismic waves from earthquakes reveal information about the structure of the Earth.

HOW DO LENSES WORK?

- There are two types of lenses: convex and concave.
- Convex lenses can converge light, bringing it to a focus, and concave lenses diverge light.
- We can draw ray diagrams to show the formation of images by lenses.
- Only convex lenses magnify.

Describing waves

Learning objectives:

- describe wave motion
- define wavelength and frequency
- apply the relationship between wavelength, frequency and wave velocity.

Light and sound are both waves, but in a thunderstorm, you see lightning before you hear thunder. Light is almost instantaneous on Earth but sound travels at about 330 m/s. How can you use this to estimate how far away a thunderstorm is?

Wavelength, amplitude and frequency

Wavelength (λ) is the distance from a point on one wave to the equivalent point on the adjacent wave. Wavelength is measured in metres (m).

The **amplitude** of a wave is the maximum displacement of a point on a wave away from its undisturbed position.

Frequency (f) is the number of complete waves passing a point in 1 second. It is measured in **hertz** (Hz). A frequency of 5 Hz means there are five complete waves passing a point in 1 second. Frequencies are also given in kilohertz (kHz) and megahertz (MHz).

1000 Hz = 1 kHz

1000 kHz = 1 MHz

Time period (T) is the time to complete one wavelength.

Time period is the reciprocal or inverse of frequency.

$$\text{period} = \frac{1}{\text{frequency}} \text{ or } T = \frac{1}{f}.$$

For example: When the frequency of a wave is 5 Hz there are five waves passing a point each second. Calculate the time period of the wave.

The period is the time for one wave to pass a point.

$$T = \frac{1}{f}$$
$$= \frac{1}{5}$$
$$= 0.2 \text{ s}$$

1. A wave has a frequency of 2 Hz. How many waves pass a point in 1 second?

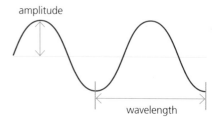

Figure 6.1 Amplitude and wavelength of a wave

REMEMBER!

Displacement includes both the distance an object moves and the direction.

DID YOU KNOW?

Sea waves carry a lot of kinetic energy (Figure 6.2). In 2014 severe storms destroyed part of the sea wall and the railway line at Dawlish in Devon.

Figure 6.2 These waves carry a lot of kinetic energy

2 Suggest how the amount of energy transferred by a wave changes as the amplitude increases.

3 Work out the frequencies of waves with time periods of:

a 0.1 s

b 0.25 s.

The wave equation

A wave transfers energy. The wave speed is the speed that the wave transfers energy across a distance, or the speed the wave moves at, in metres per second.

All waves obey the wave equation:

wave speed = frequency × wavelength

$v = f\lambda$

where the wave speed v is in metres per second (m/s), the frequency f is in hertz (Hz) and the wavelength λ is in metres (m).

If we know two of these three variables we can calculate the third using the wave equation.

Example: A radio station produces waves of frequency 200 kHz and wavelength 1500 m.

a Calculate the speed of radio waves.

b Another station produces radio waves with a frequency of 600 kHz. What is their wavelength? Assume that the speed of the wave does not change.

a $v = f\lambda = 200\ 000\ \text{Hz} \times 1500\ \text{m}$

$= 300\ 000\ 000\ \text{m/s}$

$= 3 \times 10^8\ \text{m/s}$

b Wavelength, $\lambda = \dfrac{v}{f}$

$= \dfrac{300\ 000\ 000\ \text{m/s}}{600\ 000\ \text{Hz}}$

$= 500\ \text{m}$

> **REMEMBER!**
>
> Always check that you put values into the wave equation using SI units (e.g. wavelengths in metres, frequency in hertz). If the values in the question are **not** in SI units, you first have to convert them.

DID YOU KNOW?

The light that we can see has a wavelength of about $\dfrac{1}{2000}$ of a millimetre!

4 A wave has a frequency of 2 Hz and a wavelength of 10 cm. What is the speed of the wave?

5 When the frequency of a wave doubles, what happens to its wavelength?

6 The frequency of a wave triples and its wavelength doubles. What has happened to its speed?

Transverse and longitudinal waves

Learning objectives:

- compare the motion of transverse and longitudinal waves
- explain why water waves are transverse waves
- explain why sound waves are longitudinal waves.

Why wouldn't anyone hear you if you screamed in space?

Transverse and longitudinal waves

Transverse and **longitudinal** waves can be produced on a Slinky spring.

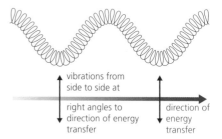

In a transverse wave on a spring the vibrations of the particles are at right angles to the direction of the energy transfer (Figure 6.3). If the particles move up and down, the energy carried in the wave is transferred horizontally, away from the energy source creating the wave. The wave moves but the spring oscillates about a fixed position.

vibrations from side to side at right angles to direction of energy transfer | direction of energy transfer

Figure 6.3 A transverse wave on a Slinky spring

① **Make a sketch of a transverse wave to explain what is meant by the terms (a) amplitude and (b) wavelength.**

Ripples on water are transverse waves (Figure 6.4). The wave looks as if it is moving outwards but the water particles actually move up and down.

A cork floating in water bobs up and down on a water wave. It only has vertical motion and it is not carried along by the wave.

② **What evidence do you have that suggests water waves are transverse waves?**

③ **Alex says, 'The particles in a transverse wave do move because waves carry items over the sea.' Suggest reasons for why he is wrong.**

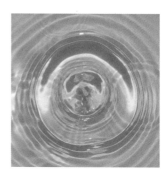

Figure 6.4 If the water moved outwards it would leave a hole in the centre

In a longitudinal wave the vibrations of the particles are parallel to the direction of energy transfer (Figure 6.5). Longitudinal waves show areas of **compression** and **rarefaction**. A compression is when the particles are closer together. A rarefaction is when they spread out. If the particles move from side to side, the energy carried in the wave moves parallel to the vibrations of the particles, away from the energy source.

Sound waves in air are longitudinal waves.

④ **When the oscillations are at right angles to the direction of energy transfer, the wave is a _____ wave.**

    A longitudinal   **B** sound   **C** standing   **D** transverse

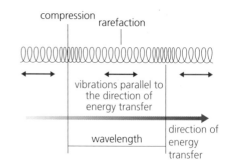

compression rarefaction

vibrations parallel to the direction of energy transfer

wavelength | direction of energy transfer

Figure 6.5 A longitudinal wave on a Slinky spring

5. Describe a method you could use to measure the amplitude of a longitudinal wave on a Slinky spring.

Sound waves

All sound is produced by vibrating particles. The vibrations are parallel to the direction of the energy transfer. The particles bunch up and spread out. This sets up a pressure wave with compressions and rarefactions.

Sound waves are longitudinal. However, a microphone connected to a cathode ray oscilloscope can be used to display sound waves on graphs of potential difference against time, so they look like transverse waves. It makes it easier for us to see and measure the frequency of sound waves.

Figure 6.7 shows two sound waves. They have the same amplitude but one wave has twice the frequency of the other wave. The frequency is found from the time for one complete cycle of the wave.

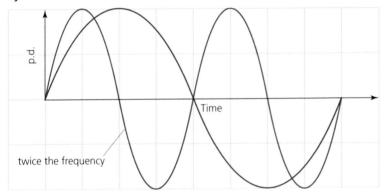

Figure 6.7 High and low frequency waves displayed on a CRO

6. Which wave in Figure 6.7 has the highest frequency?

7. Adjacent compressions in a sound wave are 15 cm apart. What is the wavelength of the sound?

Change in speed of sound waves

When a sound wave is transmitted across a boundary from one medium to another, its speed may change. When the speed of a wave changes, there is a change in the wavelength, but there is no change in the frequency. This is because the number of waves leaving the medium each second is the same as the number of waves entering the medium each second.

8. Ann puts her ear to touch an iron railing. Jack hits the iron railing with a stick about 5 m away from Ann. Explain why she hears two sounds, one after the other.

9. The speed of sound in air is 331 + 0.6T m/s, where T is the temperature in °C.

Calculate the wavelength of a 1 kHz sound wave in air at a temperature of 50 °C.

KEY INFORMATION

Sound waves cannot travel in a vacuum because longitudinal waves travel by passing the vibrations from one particle to another.

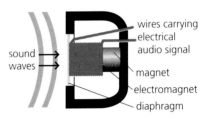

Figure 6.6 A sound wave reaching a microphone makes the diaphragm vibrate

MAKING CONNECTIONS

A microphone converts the pressure variations in sound waves into variations in electric current (see Chapter 7).

Medium	Speed of sound (m/s)
air	330
water	1500
steel	5000

KEY INFORMATION

A medium (plural media) is a material through which a wave travels. The medium itself does not travel.

KEY CONCEPT

Transferring energy or information by waves

Learning objectives:

- understand that all waves have common properties
- understand how waves can be used to carry information
- understand various applications of energy transfer by different types of electromagnetic waves.

KEY WORDS

absorb
amplitude
energy
 transfer
vibration

All waves transfer energy or information from one place to another. Water waves at the seaside transfer energy when they hit the shore to move sand and shingle up the beach. Radio waves transfer picture and sound information from the transmitter to your television set at home.

A wave is a regular vibration that carries energy. Ripples on the surface of a pond, sound in air, ultrasound, visible light, X-rays and infrared rays are all types of wave. In water, the surface just moves up and down, but the energy is carried outwards from the source. A Mexican wave in a stadium is caused by spectators just standing up and sitting down but the wave travels all round the stadium (Figure 6.8).

Why not look on YouTube at the world record of a Mexican wave?

Figure 6.8 A Mexican wave

Common properties of waves

No matter what their speed, wavelength or frequency, waves **transfer energy**. When we watch a firework display, sound waves travel slowly compared with light waves but they still transfer energy from the explosion to our eardrums. We see the flash of the explosion when light waves transfer energy to sensors in our eyes.

Amplitude measures the height of the wave above or below its rest point. The larger the amplitude of a water wave, the more energy it can transfer. An underwater earthquake can create a tsunami with waves 30 m high near the shore. This huge amplitude wave could transfer enough energy to power the whole of the UK for a year.

1. Why do you hear the sound of thunder after you see the flash of lightning?

2. Draw a diagram to show two transverse waves with different amplitudes.

3. What would you notice if the amplitude of a sound wave increased?

Using waves to transmit information

Since waves can carry energy, we use them to transmit information by varying the amount of energy carried by the wave. This can be done by simply switching the wave source on and off to create a pulsed code, as in Morse code by light, or by varying the frequency or amplitude of the wave.

Visible and infrared light is used to send internet data and telephone calls down fibre optic cables. Information that has been transformed into binary code is sent as pulses of light to be converted into digital signals read by computers or converted into vibrations in the air by telephones. The use of fibre optic cables and light waves enables vast amounts of information to be sent over far greater distances than with copper wires.

4 **What are the advantages of fibre optic cables over copper wires?**

Electromagnetic waves

Each different part of the electromagnetic spectrum is used to transfer energy.

- Microwaves can transfer data to mobile phones.
- An electric fire transfers energy to our bodies by infrared waves warming us up.
- Some energy from the Sun is transferred by ultraviolet rays.
- Not all energy from an X-ray machine is transferred. Some is absorbed by the body when an X-ray image is produced.
- Energy from radioactive sources can be transferred by gamma rays.

Mobile phones use microwaves that are similar to the waves that are used in microwave ovens (Figure 6.10). However, strict limits are applied to the amount of energy a mobile phone can transfer. In the UK it is illegal to sell phones that transfer more than 2 J of energy per second (2 W). This energy is transmitted in all directions not just into your brain. Typically your phone transfers about 5000 times less energy to your brain than a microwave oven would.

5 **For each of the examples of electromagnetic waves in the list above, suggest one piece of evidence that shows the energy transferred by the wave can be either absorbed or reflected.**

6 **If energy is being transmitted away from the Sun then why isn't the Sun continually cooling down?**

DID YOU KNOW?

Optical fibres transmit data at 200 000 km/s, which is the speed of light in glass. Many telephone conversations and computer data travel long distances through optical fibre cables with little energy loss by absorption in the glass.

Figure 6.9 Light is transmitted along fibre optic cables

Figure 6.10 Do mobile phones cook your brain?

Measuring wave speeds

Learning objectives:

- explain how the speed of sound in air can be measured
- explain how the speed of water ripples can be measured
- describe the use of echo sounding.

Mice can sing like birds, but usually at such high frequencies (up to 70 000 Hz) we don't hear them.

Measuring the speed of sound in air

Zoe and Darren measured the speed of sound in air. Sound reflects off a wall in a similar way to light reflecting off a mirror. The reflected sound is called an **echo**.

Zoe stood 50 m away from a large wall. She clapped and listened to the echo (Figure 6.11).

Zoe tried to clap each time she heard an echo while Darren timed 100 of her claps with a stop clock. He timed 100 claps in 40 s.

The time between claps is $\dfrac{40}{100} = 0.4$ s

During the time from one clap to the next the sound had time to go to the wall and back, a distance of 100 m.

$$\text{Speed} = \frac{\text{distance}}{\text{time}}$$
$$= \frac{100\,\text{m}}{0.4\,\text{s}}$$
$$= 250 \text{ m/s}$$

1 Jo and Sam also measured the speed of sound in air using the same method. They counted 50 claps in 23 s. Jo also stood 50 m from the wall.
What value did they get for the speed of sound?

2 Suggest why this method is not likely to produce an accurate value for the speed of sound in air.

The speed of water waves

The speed of a water wave can be found by measuring the time it takes for a water wave to travel a measured distance. For example, make a splash at one end of a 25 m swimming pool and measure, with a stop clock, the time it takes the wave to travel to the other end.

DID YOU KNOW?

For ripples on a water surface or sound waves in air, it is the wave and *not* the water or air itself that travels.

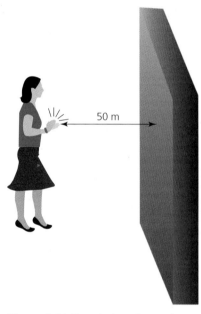

50 m

Figure 6.11 Zoe tried to clap each time she heard an echo

3 a Is the swimming pool method an accurate way of measuring the wave speed? Explain why.

b Explain how you could improve the accuracy.

4 The crest of an ocean wave moves a distance of 20 m in 10 s. Calculate the speed of the ocean wave.

Echo sounding

Many animals, such as cats and dogs, can hear sounds of a higher frequency than humans. Bats emit pulses of sound between 20 000 Hz and 100 000 Hz. They find their way around by listening to the echoes. This is echo location. We cannot hear such high frequencies.

Ships use high frequency sound waves to find the depth of the seabed or to locate a shoal of fish (Figure 6.12). This is **echo sounding**.

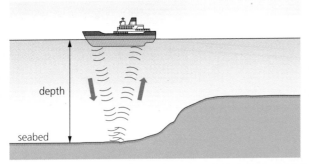

Figure 6.12 Ships use echo sounding to find the depth of the seabed

> **REMEMBER!**
> ....
> In echo sounding remember the wave goes 'there and back', so make sure you use the correct distance in calculations.

Example: A ship sends out a sound wave and receives an echo after 1 second. The speed of sound in water is 1500 m/s. How deep is the water?

Time for sound to reach the seabed = 0.5 s.

$$\text{Speed} = \frac{\text{distance}}{\text{time}}$$

distance = speed × time

$\qquad$ = 1500 m/s × 0.5 s

$\qquad$ = 750 m

5 Ships also use echo sounding to detect shoals of fish. The echo from the shoal of fish in Figure 6.13 is received after 0.1 s.

$\quad$ a $\quad$ How far below the boat is the shoal of fish?

$\quad$ b $\quad$ In Figure 6.13 the reflected pulse lasts longer than the emitted pulse. Suggest a reason for this.

6 A ship is 220 m from a large cliff when it sounds its foghorn.

$\quad$ a $\quad$ When the echo is heard on the ship, how far has the sound travelled? (The speed of sound in air is 330 m/s.)

$\quad$ b $\quad$ How long is it before the echo is heard?

7 Explain why the pulse of sound needs to be very short for accurate echo location.

Figure 6.13 Detecting fish using echo sounding

REQUIRED PRACTICAL

Measuring the wavelength, frequency and speed of waves in a ripple tank and waves in a solid

Learning objectives:

- develop techniques for making observations of waves
- select suitable apparatus to measure frequency and wavelength
- use data to answer questions.

These pages are designed to help you think about aspects of the investigation rather than to guide you through it step by step.

Frequency and wavelength of waves in a ripple tank

We can use a set of equipment called a ripple tank to explore waves. By careful observation and measurement we can measure and calculate the **wavelength** and **frequency** of the waves and then work out their **speed**. A strobe light can be used to 'freeze' the movement of the waves for making certain measurements.

1. A motor is attached to the wooden rod. What does this do?

2. What are the units of:
   a wavelength?
   b frequency?

3. Suggest what equipment you could use to measure the wavelength, and how you should set it up.

4. When measuring the wavelength, you might measure the length of 10 waves on the screen or table and then divide by 10. Explain why this is done.

5. Louise is looking at a certain point on the screen and counting how many waves pass that point in 10 seconds. How can she then calculate the frequency of the waves?

Speed of waves in a ripple tank

Sahil's group are comparing two ways of working out the speed of the waves as they travel through the water. One of these is by using speed equals distance/time and the other is by using speed equals frequency times wavelength.

6. Explain how the group could measure the speed of the waves using the equation speed = distance/time.

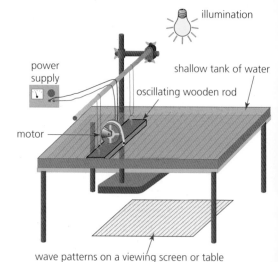

Figure 6.14 Ripple tank used for observing waves in water

Figure 6.15 Shadows of waves travelling across a ripple tank. The wavelength is the distance between two dark patches, which are the peaks (crests) of the waves

7 Now explain how the group could calculate the speed of the waves using the equation speed = frequency × wavelength.

8 Why do scientists sometimes try to use two different methods to answer a question?

Speed of sound on a stretched string

We can use a mechanical vibrator to vibrate a stretched string or elastic cord (Figure 6.16). It transfers energy continuously to the string. This produces transverse waves that travel along the string, just like when you shake a rope up and down. Changing the vibration frequency changes the wavelength of the waves. The speed of the wave travelling along the string stays the same – it depends on how heavy the string is and how tightly it is stretched.

By changing the frequency of vibration, it is possible to produce a stable wave pattern. You can then count the number of wavelengths along the string and work out their speed. The frequency will be the frequency of the signal generator.

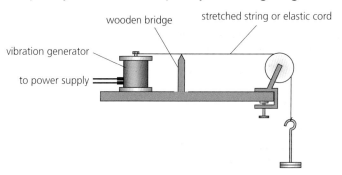

Figure 6.16 Setting up waves on a stretching string

9 Suggest what equipment you could use to measure the wavelength, and how you should set it up.

10 What difficulties are there in measuring the wavelength accurately in this experiment?

11 By changing the frequency of the vibration generator or changing the length of string allowed to vibrate (moving the wooden bridge in Figure 6.16). it is possible to increase the number of 'loops' seen in the vibrating string.

How could you use this effect to verify the speed of the waves on the string?

DID YOU KNOW?

It may seem strange that you are calculating a wave speed along the string when the string is moving up and down and has some points that don't move at all. In this special situation there are two waves travelling in opposite directions along the string (the wave created at the vibrating end and a wave reflected at the other end). These two waves meet and actually pass through each other!

Reflection and refraction of waves

KEY WORDS

absorption
normal
ray diagram
reflection
refraction
transmission

Learning objectives:

- describe reflection, transmission and absorption of waves
- construct ray diagrams to illustrate reflection
- construct ray diagrams to illustrate refraction.

When sunlight enters sea water, most of the visible light is absorbed within 10 m of the surface. Some is absorbed by phytoplankton, the producers in the marine food web. The rest of the energy that is absorbed increases the thermal energy store of the oceans, which helps drive ocean currents.

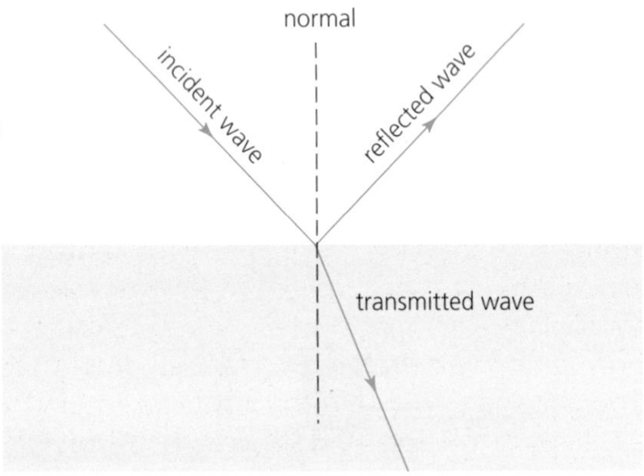

Figure 6.17 **Reflection** and **transmission** at a boundary between two materials

Reflection, absorption and transmission

Waves travel out from a point source in all directions. A **ray diagram** is a model that shows a number of lines (rays) travelling in a straight line between the wave source and an object or surface. The arrow on a ray diagram shows the direction that the wave travels. The incident ray is the ray coming in from the wave source to the surface. The reflected ray is the ray coming away from the surface.

When a wave meets the boundary between two materials, some of its energy is reflected, some is **absorbed** and some is transmitted (Figure 6.17). For example, in a thick piece of glass like a shop window you can see your reflection and see through it at the same time. In the case of a perfect reflector, no energy is transmitted or absorbed.

The dashed line in Figure 6.17 is called the **normal**. It is at 90° to the boundary and is the line from which all angles are measured.

1. **Describe what you would observe if the following waves were reflected: light, sound and water waves.**

2. **Describe what you would observe if a sound wave crossed from air into a material where it was:**
   a completely absorbed
   b partly absorbed and partly transmitted.

Law of reflection

When a wave is reflected off a surface, the angle of incidence (*i*) is equal to the angle of reflection (*r*). This is called the law of reflection. We can show this in a ray diagram for the reflection of light from a flat (plane) mirror. (Figure 6.18).

3 A ray of light hits a plane mirror at an angle of incidence of 40°. How big is the angle of reflection? Show on a ray diagram how the light is reflected.

4 The angle between a plane mirror and a ray of light is 35°. How big is the angle of reflection? Show on a ray diagram how the light is reflected.

incident ray going towards the mirror

reflected ray bouncing back from the mirror

normal

angle of incidence, *i*

angle of reflection, *r*

plane mirror

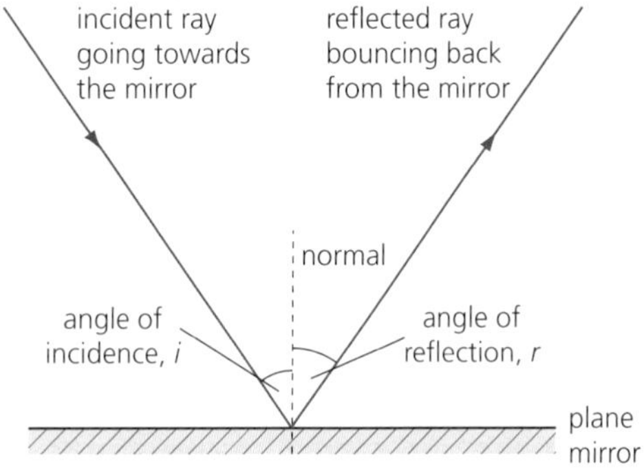

Figure 6.18 The reflection of light from a plane mirror

Refraction

A wave changes direction when it enters a different medium. This is called **refraction**. In Figure 6.17 the transmitted wave is refracted.

Figure 6.19 shows a ray of light in air, entering a glass block. The glass is denser than air, so the light slows down and bends towards the normal. The angle the refracted ray makes with the normal to the refracting surface is called the angle of refraction. You can see that when the light ray enters the glass block the angle of refraction (*r*) is smaller than the angle of incidence (*i*).

When the light ray exits the glass block, the light speeds up again and the ray refracts again, this time away from the normal. The angle of refraction is greater than the angle of incidence.

This effect can also be seen with water waves moving between deep and shallow water.

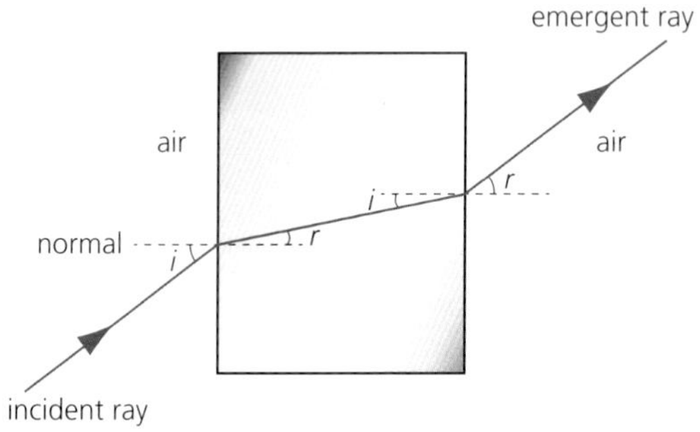

> **COMMON MISCONCEPTION**
>
> **Remember that the angles *i* and *r* are measured from the normal.**

> **DID YOU KNOW?**
>
> Light is trapped inside an optical fibre purely by the angle the light ray enters one end of the fibre. Every time the light ray reaches one side of the fibre the angle of incidence is too large for the light to escape and it is reflected back into the fibre.

emergent ray

air

air

normal

i *r*

i *r*

incident ray

Figure 6.19 Ray of light going through a glass block

5 When is the angle of refraction bigger than the angle of incidence?

6 Copy and complete the ray diagram in Figure 6.20 to show a ray of light travelling from air to water.

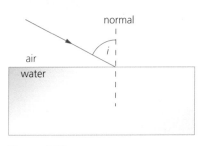

normal

i

air

water

Figure 6.20

REQUIRED PRACTICAL

Investigate the reflection of light by different types of surface and the refraction of light by different substances

Learning objectives:

*   make and record observations of how light is reflected and transmitted at different surfaces
*   measure angles and discuss the method, apparatus and uncertainty in measurements
*   draw conclusions from experimental results.

These pages are designed to help you think about aspects of the investigation rather than to guide you through it step by step.

- -

Waves can be reflected, transmitted or absorbed at the boundary between two different materials.

Reflection at different surfaces

Lots of objects reflect light. When you look around the room, most of the objects you can see are visible because they reflect light (the exceptions are ones that make their own light such as the Sun and light bulbs).

However, different objects reflect light in different ways. Some objects reflect light so as to form an image. A reflector such as a mirror forms a **specular reflection**. The surface is very smooth and all the light rays coming from one direction are reflected at the same angle.

Other objects, though, work in a different way.

You cannot see a reflection in a rough surface such as clothing or paper. At a rough surface, the light rays coming from one direction are reflected at many different angles. This is **diffuse reflection**.

Figure 6.21 Reflection in a mirror – an image is formed

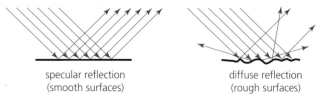

specular reflection
(smooth surfaces)

diffuse reflection
(rough surfaces)

Figure 6.22 Specular and diffuse reflection

1 Give three examples of objects that produce diffuse reflection.

2 Explain why it might be dangerous for bicycle reflectors to use specular reflectors.

DID YOU KNOW?

Retroreflectors used on bicycles and safety clothing are good at reflecting light but you can't see an image in them. All the light coming from one direction is reflected back in the same direction.

Figure 6.23

Measuring angles

Clearly, angles are important when it comes to reflection. Gemma and her group were asked to find out how the angle that light approached a mirror at would affect the angle it was reflected at. They wanted to see if there was a pattern in the results.

3 Suggest what equipment they could use, how they should set it up and what measurements to take.

4 When they are measuring the angles of the rays, where should they measure them from?

5 What difficulties are there in measuring angles accurately with this apparatus? What could you do to reduce the uncertainty in the results?

6 How would you process data from the experiment, to see if there is a relationship that links the incident ray to the reflected ray?

7 Gemma thinks that these results only apply to specular reflections – is she right?

> **REMEMBER!**
>
> When measuring the angles of light rays, remember to measure them from the **normal**, the line drawn at right angles to the surface where the ray meets the other medium.
> Remember the reflective surface of the mirror is the silvering behind the glass not at the front.

Exploring refraction

If a light ray hits another transparent medium, such as glass or water, it may travel through. Alex's group is investigating **refraction**; they used a glass block and a single ray of light. Figure 6.24 shows what they found from one of their experiments, by using a sharp pencil to mark the centre of each ray.

8 What conclusion can they form from this data about what happens to the ray of light when it:

   a enters the glass block?
   b leaves the glass block?

9 What should they notice about the direction of the ray leaving the block compared with the direction of the ray entering the block?

10 Do you think they would get similar results no matter what angle the light approached the block at?

11 Describe how Alex's group could investigate the refraction of light in water. What should the container be like?

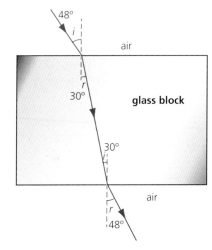

Figure 6.24 Ray diagram showing the results from one of the experiments of Alex's group

Sound waves

Learning objectives:

- describe how we hear sound and state the range of frequencies we can hear
- explain that sound travels faster in a denser medium.

KEY WORDS

absorption
eardrum
reflection
sound wave
transmission

Bats hear sound of frequencies from 30 Hz to 100 000 Hz, so they can hear much higher-pitched notes than humans.

HIGHER TIER ONLY

Range of hearing

The typical range for human hearing is between 20 Hz and 20 kHz (20 000 Hz). Young children can hear a greater range of frequencies than adults, but the range decreases rapidly as you get older.

The range of frequencies a person can hear is reduced if they are exposed to loud sounds for a long period.

1 Could a person hear the following frequencies?

  a   40 000 Hz

  b   4000 Hz

The speed of sound in different media

Solids are denser than liquids and gases. Because the particles are packed together tightly, sound travels fastest in solids. Liquids are denser than gases, but less dense than solids. Sound travels most slowly in gases because they are the least dense. The particles in gases are very far apart compared with solids and liquids.

2 Explain why sound travels faster in denser materials, such as steel, than in air.

Hearing sounds

Sound is a longitudinal wave. The particles vibrate along the direction of the wave. This is a pressure wave with compressions and rarefactions (Figure 6.25).

Sound usually reaches the ear as vibrations in the air. We can hear because the kinetic energy of the sound wave in air is transferred to the **eardrum** and then to other parts of the ear, and then electrically to the brain.

Sound waves travel through solids as vibrations. The eardrum is taut skin; sound waves entering the ear make the eardrum vibrate and this transfers vibrations to tiny bones behind it.

DID YOU KNOW?

Sound can travel much further and much quicker underwater than in air. Blue whales call to each other over distances of 1000 km.

These vibrations have a frequency – the number of vibrations per second. This works best over a certain frequency range, which restricts the limits of human hearing.

Sound travels to the ear as a pressure wave - a continuous variation of high and low pressure – and exerts a varying force on the eardrum, causing it to vibrate in and out.

The same process is used in a microphone, in which the pressure wave makes the diaphragm vibrate. A coil is attached to the diaphragm and moves near a magnet, making a current flow.

3 If the wavelength of a sound wave increased how would it sound different?

4 Describe how the air particles are moving in Figure 6.25.

5 Use the diagram of the ear to explain how energy is transferred from the air to nervous impulses.

6 If someone sings with a loud high pitched note into a microphone, what effect will this have on the electrical impulses produced in the coil?

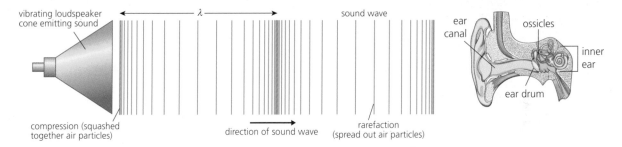

Figure 6.25 A sound wave showing compressions and rarefactions

Exploring ultrasound

Learning objectives:

- explain what ultrasound is
- describe how ultrasound is used in medicine and in industry
- explain reflection, absorption and transmission of sound.

KEY WORD

ultrasound

Ultrasound scans can be used to see inside objects without damaging them.

HIGHER TIER ONLY

What is ultrasound?

Ultrasound is sound waves with a frequency above 20 kHz. When a sound wave meets a boundary between two materials, such as between different tissues or organs, some sound is always absorbed, some is reflected and the rest is transmitted through the boundary. The strongest reflection (echo) occurs when the sound wave hits something where the speed of sound is very different. There is then less sound to be absorbed or transmitted. This is how an ultrasound scan works.

DID YOU KNOW?

Ultrasonic test trains are used on the move to test for internal cracks in train rails which could fracture and cause possible derailment.

1. How is ultrasound different from sound?

2. Explain why normal sound wouldn't produce a clear image.

Uses of ultrasound scans

Ultrasound scanning is used to build up an image of an internal structure that cannot be seen from outside. It is used in industry:

- to find cracks or gaps in metal objects such as aircraft structures and oil pipelines
- to measure the thickness of objects due to ultrasound being reflected from both surfaces.

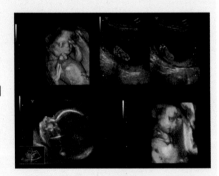

Figure 6.26 A 3-D ultrasound scan of a fetus at about 25 weeks gestation

Ultrasound scans are used in medicine because ultrasound waves, unlike X-rays, have low energy and do not damage living cells. Medical ultrasound scans are used:

- to check that an unborn baby is developing properly by creating an image of the fetus from ultrasound reflected from different parts of the fetus
- to measure the speed of blood flow in a vein or artery.

3. What are the advantages to patients of using ultrasound instead of X-rays to image developing babies?

4. Give two uses of ultrasound scanning.

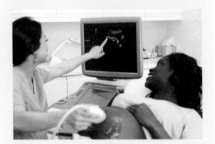

Figure 6.27 An ultrasound scan being carried out on a pregnant woman

5 Explain why ultrasound is useful when engineers suspect a buried oil pipeline has a leak.

How ultrasound scans work

When a patient has an ultrasound scan (Figure 6.28), a gel is placed on their body between the ultrasound probe and their skin. Without gel, nearly all the ultrasound would be reflected at the skin and a good image of the internal structure would not be obtained.

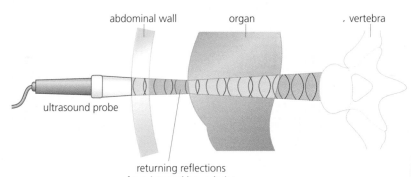

abdominal wall   organ   vertebra

ultrasound probe

returning reflections
from internal boundaries

Figure 6.28 How ultrasound is reflected during a body scan

When a pulse of ultrasound waves passes through lots of different materials the returning echoes can be recorded. Reflections from different layers return at different times from different depths. The depth of each structure is calculated using speed = distance/time. The data is used to build up an image of the different layers in the internal structure.

6 The speed of sound is very different in blood and bone. How does this affect the amount of reflection that occurs? Explain why this might be a problem.

7 Why is a gel used when a person has an ultrasound scan?

8 Describe how an ultrasound scan image is produced.

9 The difference in time between the return of ultrasound pulses from either side of the head of a fetus is 140 µs (140 × 10^{-6} s). If the speed of ultrasound in the head is 1500 m/s, how big is its head?

10 Explain why it would be impossible to produce an ultrasound image of an object behind the lungs.

REMEMBER!

Remember that in an ultrasound scan the reflected ultrasound goes 'there and back', doubling the distance the ultrasound travels.

Figure 6.29 Images showing the flow of blood through the heart, taken from a Doppler ultrasound scanner. The speed of blood changes the frequency of the ultrasound, which can be detected by the scanner

Seismic waves

Learning objectives:

- describe how earthquakes are detected
- describe the properties of P-waves and S-waves
- explain how the properties of seismic waves allow us to investigate the inside of the Earth.

KEY WORDS

earthquakes
P-waves
S-waves
seismic waves

Earthquakes happen more often than you think. The UK is actually hit by hundreds of small earthquakes every year, with the largest on record measuring 6.1 on the Richter scale.

HIGHER TIER ONLY

Detecting earthquakes

Earthquakes happen when two parts of the Earth's crust slide past each other suddenly at a fault. Shock waves called **seismic waves** pass through the Earth and also travel around its surface. It is the surface waves that cause damage to houses and other structures. We use a seismometer to detect seismic waves from an earthquake or explosion (Figure 6.30).

1. What force attracts the pen of a seismometer towards the centre of the Earth?

2. Suggest why the base of a seismometer is bolted to solid rock.

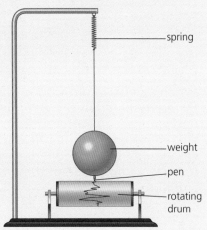

Figure 6.30 A seismometer detects the arrival of seismic waves

P- and S-waves

An earthquake happens below the Earth's surface at the focus. The epicentre is the point on the Earth's surface above the focus.

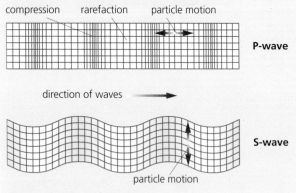

Figure 6.31 P- and S-waves

There are two main types of seismic waves (Figure 6.31):

- A **P-wave** is a primary (pressure) wave. It is a longitudinal wave similar to a sound wave. The speed of P-waves generally increases with depth in the Earth, but is slower in liquids than in solids.

- An **S-wave** is a secondary (shear) wave. It is a transverse wave. The speed of S-waves generally increases with depth in the Earth, but is always less than P-waves. S-waves cannot travel through liquids.

3 What type of wave is (a) a P-wave? (b) an S-wave?

4 Which type of wave travels faster, P or S?

Evidence for the structure and size of the Earth's core

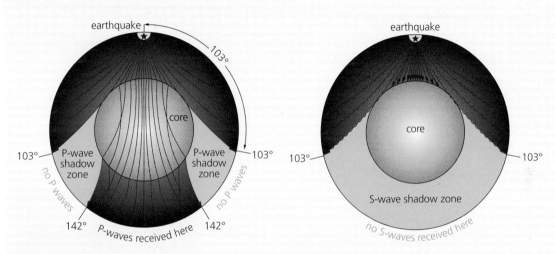

Figure 6.32 P-waves

Figure 6.33 S-waves shadow zone

- P- and S-waves spread out in all directions from an earthquake. When the waves reach the Earth's surface they are detected by seismometers. There are two narrow 'shadow zones' on the Earth's surface at certain distances from the earthquake where no P-waves are detected (Figure 6.32). Since P-waves can travel through both liquids and solids, P-waves must be refracted away by entering a region with lower P-wave velocity. P-waves are detected directly on the opposite side of the Earth to an earthquake, showing that P-waves are refracted back when they reach a region of higher velocity.
- There is a larger S-wave shadow zone on the opposite side of the Earth to an earthquake, where no S-waves are detected (Figure 6.33). Since S-waves can pass through solids but they cannot travel through liquids, part of the centre of the Earth must be liquid.
- By analysing the size of the P- and S-wave shadow zones from several earthquakes, it is possible to work out the depth within the Earth at which a liquid area slows down P-waves and stops S-waves. This is the boundary between the lower mantle (solid) and the outer core (liquid) at a depth of 2900 km.

5 Explain why the paths of the P- and S-waves are curved when they move towards and away from the core.

The electromagnetic spectrum

Learning objectives:

- recognise that electromagnetic waves are transverse waves
- describe the main groupings and wavelength ranges of the electromagnetic spectrum.

KEY WORDS

electromagnetic waves
electromagnetic
 spectrum
longitudinal wave
transverse wave
visible spectrum

All electromagnetic radiation from the Sun (including visible light) travels the 149 million kilometres to the Earth in about 8 minutes.

Electromagnetic waves

We know light is a wave because it has the same properties as other waves: it can be reflected and refracted.

White light is a mixture of waves with different wavelengths which we see as different colours. These colours can be separated into what we call the **visible spectrum**. When scientists investigate the visible spectrum they can detect invisible waves on both sides, showing that the visible spectrum is really part of a much wider spectrum, which we call the **electromagnetic spectrum**.

All the waves in the electromagnetic spectrum are **transverse waves** with many properties in common with visible light.

Just like other waves, all **electromagnetic waves** transfer energy from one point to another. In electromagnetic waves, electromagnetic fluctuations occur at right angles to the direction in which energy is being transferred by the wave.

Some waves have to travel through a material. Sound waves can travel through air, liquids and solids but not a vacuum. Water ripples travel along the surface of water.

Electromagnetic waves are different from other waves because they do not need a material. They can travel through a vacuum. This is a special property of electromagnetic waves, which enables light and infrared waves to reach us from the Sun. All electromagnetic waves travel at the same speed in a vacuum, 3.0×10^8 m/s.

KEY INFORMATION

Visible light is electromagnetic radiation that our eyes can detect. This is just a limited range of the electromagnetic spectrum.

1. What are the similarities and differences between transverse and longitudinal waves?

2. Explain how waves in the electromagnetic spectrum are different from other waves.

3. What properties do all electromagnetic waves have in common?

The electromagnetic spectrum

Figure 6.34 shows that electromagnetic waves span a wide, continuous range of wavelengths and frequencies.

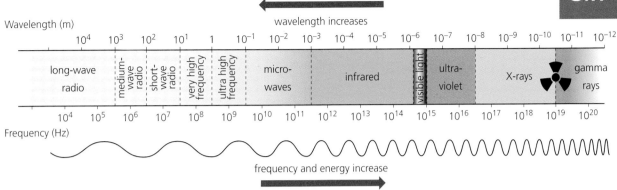

Figure 6.34 The wavelengths and frequencies of waves in the electromagnetic spectrum

The wavelength ranges of the groups of waves in the electromagnetic spectrum are (from long to short):

- radio, TV and microwaves
- infrared
- visible (red to violet)
- ultraviolet
- X-ray
- gamma ray

The shorter the wavelength of the electromagnetic wave, the further it can travel through other materials. The higher the frequency of a wave, the more energy it can transfer to another object when the radiation is absorbed.

REMEMBER!

Remember the order of electromagnetic wave groupings.

4. **Which grouping in the electromagnetic spectrum has the highest frequency?**

5. **Ultraviolet light used in a sunbed has a wavelength of 3.5×10^{-7} m. Calculate the frequency of this light.**

6. **Calculate the frequency of an electromagnetic wave with a wavelength of 20 cm. Use standard form for your answer.**

7. **Suggest why ultraviolet waves are more dangerous than radio waves.**

KEY INFORMATION

The shorter the wavelength (the higher the frequency) of an electromagnetic wave, the more dangerous the radiation.

Waves as energy carriers

Waves can transfer energy from one place to another. Heat lamps keep newly hatched chicks warm, bulbs transfer energy as light throughout a room and sunbathers receive energy directly from the sun by UV waves.

The energy becomes dissipated, however; the waves often spread out as they travel, meaning that the energy is spread over a larger area.

KEY INFORMATION

All electromagnetic waves travel at the same speed in a vacuum, 3.0×10^8 m/s.

8. **Explain how a microwave oven demonstrates the idea of waves carrying energy.**

9. **Why might it be difficult to use a mobile phone if the nearest phone mast is a long distance away?**

Reflection, refraction and wavefronts

Learning objectives:

- explain reflection and refraction and how these may vary with wavelength
- use wavefront diagrams to explain refraction in terms of the difference in velocity of the waves in different substances.

KEY WORDS

absorption
reflection
refraction
transmission
wavefront

A rainbow is caused by the refraction of sunlight by raindrops. When entering a raindrop, blue light in sunlight is slowed down more than red light, so is refracted by a greater angle.

HIGHER TIER ONLY

Reflection of electromagnetic waves

Studying how visible light behaves is relatively easy because we can see what happens. This then enables us to predict, check and understand how other electromagnetic waves behave.

All waves, including all electromagnetic waves, can be **reflected**, **transmitted** or **absorbed**.

One way of keeping things warm is to use aluminium foil to reflect the infrared radiation emitted by a warm object back towards the object. Figure 6.35 illustrates how this can be done. It can also be used to keep objects cool.

The proportion of a wave's energy that is reflected, transmitted or absorbed depends on its wavelength and on the medium it enters. For example, while some radio waves from stars pass through the atmosphere and can be detected on Earth, some radio wavelengths are reflected back into space from the Earth's upper atmosphere (the ionosphere). At other wavelengths, gas molecules in the air absorb incoming radio waves.

3. **10 MHz radio waves will pass through the ionosphere without being reflected. Calculate the wavelength of these waves.**

Refraction of electromagnetic waves

We saw earlier that when light waves travel from one medium into another, they change speed and may change direction (Figure 6.36). This is called **refraction**.

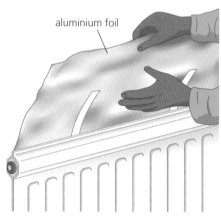

aluminium foil

Figure 6.35 Putting aluminum foil behind a radiator means the foil will reflect infrared radiation back into the room

1. **Explain how using aluminium foil behind a radiator could reduce the household energy bill.**

2. **Give two other examples where infrared radiation is reflected to keep things warm or cool.**

DID YOU KNOW?

Some telephone signals between the UK and the USA are sent by microwaves reflected from satellites.

4. **The shortest wavelength radio wave that will pass through the atmosphere without being absorbed is 100 m. Calculate the frequency of these waves.**

All electromagnetic waves can be refracted when they enter a medium in which the wave velocity is different. The shorter the wavelength, the more the wave is refracted.

Different substances refract the same wavelength in different ways. For example, radio waves travel in a straight line from the transmitter, but are refracted in the lower layer of the atmosphere. The amount of refraction can be affected by differences in atmospheric temperature and pressure.

5 **Suggest why radio signals can sometimes travel further between two points on the Earth as atmospheric conditions change.**

Explaining refraction

A **wavefront** is a line that joins all the points on a wave which are moving up and down together at the same time. The wavefront is at right angles to the direction the wave is travelling. We can see what happens when waves are refracted by looking at plane water waves in a ripple tank (Figure 6.37).

When a wave strikes a boundary at an angle, one part of the wave reaches the boundary before the rest of the wave. This part changes speed first. In Figure 6.38 the left-hand parts of the wave fronts get closer together, because waves travel slower in shallower water. The wave front changes direction. The wave is refracted towards the normal.

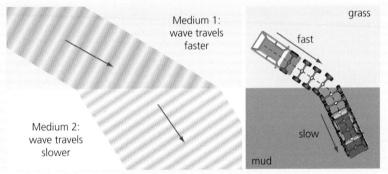

Figure 6.38 The way that waves refract when going from a higher speed medium to a lower speed medium is similar to what happens when a car drives at an angle into mud. The wheels that reach the mud first slow down first so the car changes direction

6 **Water waves are affected by the depth of the water they travel through.**

   a **What happens to the wavefronts in Figure 6.37 when the water waves move from deep to shallow water?**

   b **Do the waves travel faster in deep or shallow water? Explain your answer.**

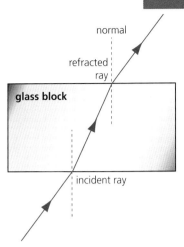

Figure 6.36 Light is refracted towards the normal when it travels from air to glass, and away from the normal when it travels from glass to air

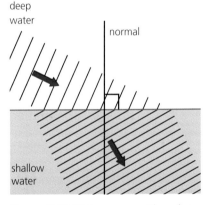

Figure 6.37 Water waves with a plane (straight) wave front travelling from deep to shallow water

KEY INFORMATION

The wave fronts get closer together because speed = frequency × wavelength and the frequency of the waves (which depends on the source of the waves) is constant. If the speed decreases, the wavelength must also decrease.

Gamma rays and X-rays

Learning objectives:

- list the properties of gamma rays and X-rays
- compare gamma rays and X-rays.

KEY WORDS

gamma ray
radiation dose
tracer
X-ray

X-ray machines called pedoscopes were introduced in the 1930s in shoe shops in the UK. They enabled parents and children to see how well shoes fitted. They were very popular with children, who loved to watch their bones move as they wriggled their toes. Unfortunately people using these machines did not realise how dangerous X-rays can be, and some shoe shop assistants might have developed cancer because of using them.

Gamma rays

Gamma rays have the shortest wavelengths and transfer the most energy of all the waves in the electromagnetic spectrum. This means they can be harmful to living cells. Gamma rays are used to kill cancer cells or reduce the size of a tumour (radiotherapy). However, the **radiation dose** from a single treatment is usually low in comparison with the level of background radiation. Radiation dose is measured in sieverts (Sv) and millisieverts (mSv).

Gamma rays are also used in medical imaging. Technetium-99m is a radioactive isotope that emits gamma rays. It is used as a radioactive **tracer**. A gamma camera monitors where gamma rays are emitted from in the body, to produce a moving image that shows how organs are functioning (Figure 6.39). Technetium-99m has a half-life (the average time it takes for half the nuclei present to decay) of 6 hours, so it stays in the body long enough for diagnosis but not long enough to cause lasting damage to cells.

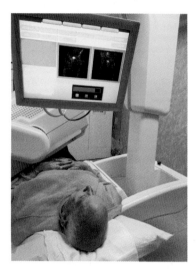

Figure 6.39 This patient is having a scan using a gamma tracer and gamma camera

MAKING CONNECTIONS

The uses of gamma rays in tracers and in radiotherapy, and how the benefits must be balanced against any potential risk, are also discussed in Chapter 4, Atomic structure.

1. Why is a half-life of 6 hours good for diagnostic scans?

2. Name other properties that a radioisotope should have when used for medical diagnosis.

X-rays

X-rays pass through soft tissues in the body but will be absorbed by bone. X-ray images can be used to check for broken bones (Figure 6.40).

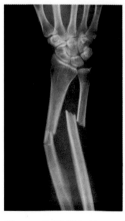

Figure 6.40 This X-ray shows a broken ulna in the wrist

X-rays are also used in computerised tomography (CT) scans. Multiple images are taken at many different angles to build up a detailed picture of inside a patient's body (Figure 6.41).

In addition to medical imaging for diagnosis, X-rays are used in radiotherapy to treat cancer.

3 Should X-rays used in radiotherapy have longer or shorter wavelengths than those used for medical diagnosis? Explain your answer.

4 Suggest why gamma rays are not used for medical imaging in the same way as X-rays.

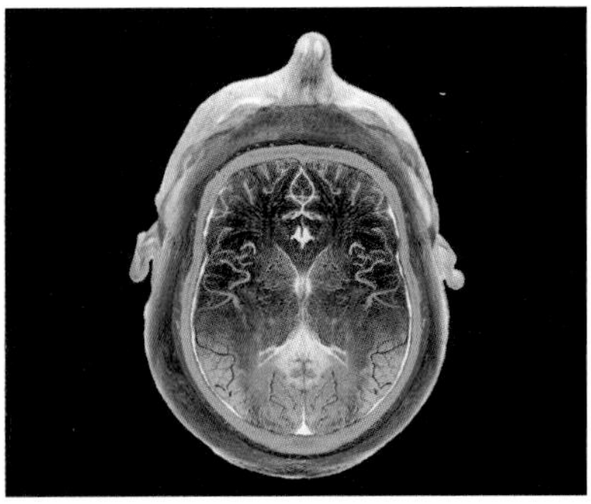

Comparing gamma rays and X-rays

Gamma rays and X-rays have very similar ranges of wavelengths and frequencies, and so have very similar effects. Their high frequencies mean they carry the most energy and are the most penetrating forms of electromagnetic radiation, so therefore they can also be the most dangerous to human body tissue when the radiation is absorbed.

Both gamma rays and X-rays are highly ionising radiation. This is how they damage and destroy cancer cells. Large doses of ionising radiation can cause the mutation of genes in cells, which can cause cells to become cancerous.

The key difference between gamma rays and X-rays is how they are produced. Gamma rays are emitted from the nucleus of an unstable atom during radioactive decay. This is a random process. X-rays are generated by an X-ray machine when high-speed electrons collide with metals and lose energy.

5 Describe the similarities and differences between gamma rays and X-rays.

6 Explain why an X-ray source needs an electric power supply but a gamma source does not.

7 Why must great care be taken when using X-rays and gamma rays?

8 Suggest how X-rays are used to detect faults in welds.

9 The radiation dose from a chest X-ray is 0.014 mSv. The dose from a CT scan of the head 1.4 mSv. The average annual dose a person in the UK receives from natural sources is 2.7 mSv, although the typical exposure for radiation workers is 5 mSv per year.
Evaluate the risks of undergoing these medical imaging tests.

Figure 6.41 A CT scan can give good pictures of soft tissue regions but exposes a patient to a much higher radiation dose than a single X-ray

6.13

KEY INFORMATION

Like all electromagnetic waves, gamma rays and X-rays travel at the same velocity in both a vacuum and air, but at slower speeds when they travel in a material.

Google search: 'using X-rays for medical imaging and treatment, technetium-99m' 217

Ultraviolet and infrared radiation

Learning objectives:

- describe the properties of ultraviolet and infrared radiation
- describe some uses and hazards of ultraviolet radiation
- describe some uses of infrared radiation.

KEY WORDS

infrared radiation
ultraviolet radiation

Some of the dyes in inks are only visible under ultraviolet light. These are used in security pens used to mark valuable equipment such as computers and bicycles. Many banknotes have a special feature which is only visible under ultraviolet light.

Ultraviolet radiation and its uses

Ultraviolet rays have shorter wavelengths than violet light. They are emitted from very hot objects (4000 °C or more) such as the Sun.

One use of ultraviolet radiation is in fluorescent lighting. This type of light is more energy-efficient than traditional filament light bulbs and this technology is used in compact fluorescent light bulbs. Fluorescent lights have a white phosphor coating inside the glass. When the UV rays strike this, it glows and emits visible light.

Our skin absorbs ultraviolet radiation from sunlight. Small doses of ultraviolet rays are good for us, as this produces vitamin D in our skin. A sun tan (from sunlight or a sunbed) is a natural darkening of the skin to protect itself from too much ultraviolet radiation.

Figure 6.42 Some of the ink used in Euro banknotes is only visible in UV light

1 State three uses of ultraviolet radiation.

2 Describe how ultraviolet radiation can be used to mark a TV.

3 Suggest why many shops have an ultraviolet lamp the till.

4 John says that ultraviolet light has a purple colour. Explain whether John is correct.

The hazards of ultraviolet radiation

Ultraviolet radiation has a higher frequency than visible light and transfers more energy when it is absorbed by skin. Too much sunlight or use of sunbeds causes premature aging of the skin, such as wrinkles and dark pigmentation spots. Large doses of ultraviolet radiation can be harmful to our eyes and may also cause skin cancer, especially in people with fair skin.

Sunscreens can reduce the risks of sunburn and skin cancer (Figure 6.43).

Figure 6.43 Sunscreens contain substances that absorb or reflect some of the Sun's ultraviolet radiation

5 Explain why small doses of ultraviolet rays can be good for you but large doses can be harmful.

6 Suggest why skiers might be at a greater risk of sunburn than people on the beach.

Infrared radiation

Infrared radiation is next to red light on the electromagnetic spectrum, and has longer wavelengths than red light. Anything that is warmer than its surroundings emits energy by giving out infrared radiation. You cannot see infrared radiation but you can feel the infrared radiation given out by hot things such as a fire, a heater or an oven.

The traditional way of cooking is by using infrared radiation. However, infrared radiation only heats the surface of food. The radiation is not transmitted very far into food. Meat that appears to be cooked may be raw on the inside (Figure 6.44).

Energy from infrared radiation is absorbed by the particles on the surface of the food. They vibrate more and energy can then be transferred slowly by conduction to the food below the surface.

7 Why is the chicken in Figure 6.44 still undercooked in the middle when it appears to be fully cooked on the outside?

8 What energy transfer processes heats the centre of the food?

Remote controls for electronic devices such as TVs and radios work by emitting infrared radiation. The remotely controlled device won't work if an object is between it and the control because the infrared radiation is absorbed. Infrared beams can also be used for security alarms.

Thermal imaging cameras (Figure 6.45) detect low levels of infrared radiation from warm objects.

9 Suggest two ways that infrared sensors can be used in a burglar alarm.

10 Suggest one way a TV remote control could still work when not pointed directly at the TV set.

DID YOU KNOW?

Halogen hobs work by using ring-shaped halogen lamps beneath a glass cooktop. Although you see a bright red light, the glowing filament in the lamp radiates mostly infrared radiation.

Figure 6.44 This chicken is under-cooked in the middle

Figure 6.45 A thermal imaging camera detects infrared radiation given off by an object or person

REQUIRED PRACTICAL

Investigate how the amount of infrared radiation absorbed or radiated by a surface depends on the nature of that surface

Learning objectives:

- explain reasons for the equipment used to carry out an investigation
- explain the rationale for carrying out an investigation
- apply ideas from an investigation to a range of practical contexts.

Anything that is warm gives off infrared radiation and the warmer it is the better an emitter it is. However, the colour of an object also affects how much radiation it emits. If we want a radiator to work as well as possible, the colour matters.

Colour also makes a difference to absorbing infrared radiation. If you want to make sure you stay cool on a hot day, choose the right clothing. You'll cook in the wrong colour!

> These pages are designed to help you think about aspects of the investigation rather than to guide you through it step by step.

Investigating absorption

Alex's teacher is showing the class how to compare different surfaces to see which is better at absorbing **infrared radiation**. She has set up two metal plates, one on either side of a heater. One plate has a shiny surface and the other has been blackened. On the back of each plate a glass stopper has been stuck on with wax. The heater is turned on. After a few minutes one of the stoppers drops off and the other follows several minutes later.

1 What does this experiment show?

2 With what you have learned about absorbers of infrared radiation, which stopper would drop off first?

3 What needs to be done to make sure the experiment is a fair test?

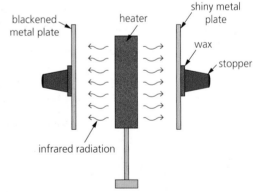

Figure 6.46 Experiment to compare absorption of infrared radiation by different surfaces

Commenting on the design of an experiment

Experiments have to be designed well if the results are to be meaningful. The experiment shown in Figure 6.47 has been designed to compare how good different surfaces are at emitting infrared radiation (radiating thermal energy). The cube is filled with very hot water and an infrared detector is held opposite each of the various faces in turn. The different faces have different finishes on them and it is possible to compare the readings from each.

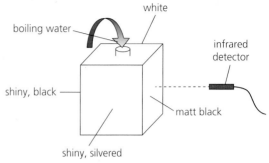

Figure 6.47 Experiment to compare infrared radiation emitted by different surfaces. The matt black surface produces the highest reading and the shiny, silvered surface produces the lowest reading

4 Why does the experiment use one can with four different surfaces rather than four cans, each a different type of surface?

5 Why is it designed so that each face has the same width?

6 What needs to be true about the positioning of the infrared detector?

An alternative way of investigating the amount of infrared radiation emitted by different surfaces is to replace the infrared detector with a thermometer with a blackened bulb.

7 Explain why a thermometer with a blackened bulb is used.

8 Which apparatus would give the greatest resolution for temperature readings – the infrared sensor or the thermometer?

DID YOU KNOW?

Firefighters use infrared cameras to locate unconscious bodies in smoke-filled buildings. Even though visible light can't penetrate the smoke to enable the body to be seen, the infrared radiation from the body can.

Applying the ideas

Some ideas in science have an immediate application and ones about infrared radiation affect our everyday lives. We can use the results of these experiments to inform us about the way that various objects should be designed.

Think about the investigation into absorbing infrared radiation.

9 What does this suggest about the best colour for:
a a firefighter's suit?
b solar panels on the roof of a house, absorbing thermal energy into water to use inside?

Now think about the experiment on emitting infrared radiation.

10 What does this suggest about the best colour for:
a a teapot?
b the pipes on the back of a refrigerator, which need to radiate thermal energy from the inside of the fridge to its surroundings?

11 Explain what you would see if you touched a thermometer to each side of the cube in Figure 6.47.

REMEMBER!

Make sure you understand about the main features of the design of an experiment. The equipment has been developed in a particular way for good reasons and you need to know what they are.

Microwaves

Learning objectives:

- list some properties of microwaves
- describe how microwaves are used for cooking food and in radar.

KEY WORD

microwaves

Mobile phones use microwaves. A mobile phone today can have more computing power than the first computers used to land man on the Moon.

Properties of microwaves

Microwaves are radio waves with short wavelengths. The wavelengths of microwaves vary between 1 mm and 30 cm. Microwaves used for communication have a longer wavelength than those used for cooking. This means that there is less energy associated with mobile phones than with microwave ovens.

1 List two uses of microwaves.

2 Suggest what use is likely to be made of microwaves with a wavelength of 25 cm.

> **KEY INFORMATION**
>
> Remember, the higher the frequency of electromagnetic waves the more energy associated with the radiation.

Cooking with microwaves

Microwave ovens do not use a flame or heated metal to cook food. They use microwaves instead. Microwave ovens cook food faster and are less expensive to run than conventional ovens.

Microwaves penetrate about 1 cm into the outer layers of food before being absorbed. The energy transferred by the microwaves makes water or fat molecules in the outer layers of food vibrate more (Figure 6.48). Energy is then transferred from the vibrating water or fat molecules to the centre of the food by conduction.

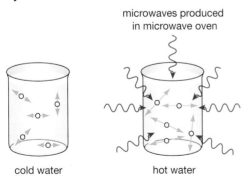

microwaves produced
in microwave oven

cold water   hot water

Figure 6.48 Microwaves cook food by making the water or fat molecules vibrate more

> **DID YOU KNOW?**
>
> The inside walls of microwave ovens must be made of metal. Metal reflects microwaves so the microwaves are trapped inside. This is important to make sure we do not cook ourselves. The door of a microwave oven is made from special glass that also reflects microwave radiation.

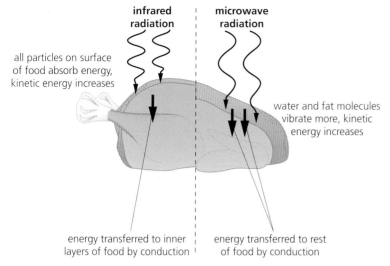

Figure 6.49 Microwaves are transmitted further into the food than infrared radiation before being absorbed

3 Explain the differences between the ways a microwave oven and a conventional oven cook food.

4 Suggest two advantages of using a microwave oven instead of an infrared oven.

5 Explain what would happen if you tried to make toast in a microwave oven.

Radar

Radar was developed during the Second World War to detect aircraft outside the normal range of sight and hearing. Radar uses microwaves, which pass through the atmosphere even when visible light is blocked by clouds or dust.

Radar works by sending out microwave pulses from a transmitter mounted on the aircraft. Microwaves reflect from a metal target in a similar way to light reflecting off a mirror. The reflected microwaves are detected by the radar system's receiving antenna.

Example: An aircraft sends out a very brief microwave pulse and receives an echo after 10.8 μs. If the speed of microwaves in air is 3.0×10^8 m/s, how far away is the target?

$$\text{Speed} = \frac{\text{distance}}{\text{time}}$$

distance = speed × time = 3.0×10^8 m/s × 10.8 μs = 3240 m

The distance to the target is half this, 1620 m.

6 If a radar echo takes 64 μs to return, how far away is the target?

7 Suggest why there is a maximum range at which targets can be detected using radar.

DID YOU KNOW?

Many people think that eating carrots helps you to see in the dark. This myth was introduced as propaganda during the Second World War, to explain the fact that pilots were able to detect enemy aircraft which were not visible at night. This really had nothing to do with carrots – they used radar to detect other aircraft in the sky.

MAKING CONNECTIONS

You do not need to remember how radar works, but you should notice that echo sounding and radar work in the same way. A signal is transmitted, and the reflected signal is used to work out the distance of the object.

KEY INFORMATION

'Micro' is a standard prefix when showing quantities. It means one millionth and is represented by the symbol μ. One microsecond is written 1μs and is 1×10^{-6} seconds.

Radio and microwave communication

Learning objectives:

- describe how radio waves are used for television and radio communications
- describe how microwaves are used in satellite communications
- describe how radio waves are produced and received
- describe the reflection and refraction of radio waves.

KEY WORDS

microwaves
radio waves
receiver
satellite
transmitter

Bluetooth is a wireless technology for exchanging data over short distances using microwaves.

Using radio waves for communication

A system using electromagnetic waves to communicate must contain a **transmitter** to send a signal and a **receiver** to receive it. Terrestrial radio and TV signals are sent by **radio waves**. The radio waves travel through the air at the speed of light. The path between transmitter and receiver usually has to be a straight line with no large obstructions such as hills and large buildings, so the transmitters are placed on towers that may be hundreds of metres tall.

1. Describe the properties of radio waves that make them useful for radio communication.

2. A radio wave has a wavelength of 100 m. What is its frequency?

HIGHER TIER ONLY

Radio waves can be produced by oscillations in electrical circuits. When a current flows through a wire it creates an electric field around the wire. When the current changes, the electric field changes. The changing current produces radio waves. This is how radio transmitters work.

Radio waves can induce oscillations in an electrical circuit, with the same frequency as the radio wave itself.

3. Suggest how a radio receiver works.

4. Some radios do not need a power supply (even a clockwork one). Suggest where they might get the energy from that they need to produce the sound.

Using microwaves for communication

Microwaves are used to transmit mobile phone signals. Microwave transmitters or base stations are placed on high buildings or masts to give better line of sight communication over large distances. Although most microwave signals are sent directly, some are sent from a transmitter to a receiver via a **satellite** (Figure 6.50). Microwave satellite communications are used for satellite phones and for satellite TV.

5 Suggest why satellite dishes are placed on the walls or roofs of houses.

6 Suggest why aerials for mobile phone signals are placed close together in towns and cities.

HIGHER TIER ONLY

Reflection and refraction of radio signals

Like all other electromagnetic waves, radio waves can be reflected and refracted. Radio waves are refracted in the upper layers of the atmosphere, called the ionosphere. The amount of refraction depends on the frequency of the wave.

Waves with a long wavelength and low frequency undergo most refraction in the ionosphere. Figure 6.51 shows how radio waves are refracted in the ionosphere so the wave returns to the Earth's surface. Microwaves with shorter wavelengths are not refracted or reflected and pass straight through the ionosphere. This is why microwaves, not radio waves, are used for satellite communication.

7 State what happens to some radio waves in the ionosphere.

8 A radio station broadcasts with a frequency of 103.4 MHz from the transmitter in Figure 6.51. Explain why a radio at point X will not receive the signal.

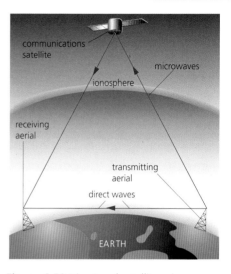

Figure 6.50 Direct and satellite microwave communication

DID YOU KNOW?

Using a satellite for communication delays a signal by less than 0.3 s.

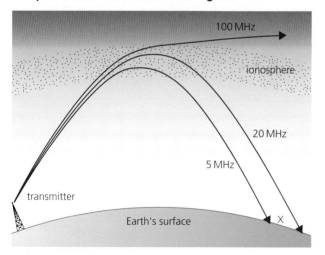

Figure 6.51 Waves of different frequencies are refracted by different amounts when entering the ionosphere

Colour

Learning objectives:

- describe what happens when light of different wavelengths lands on an object
- explain what determines the colour of an opaque object
- explain the effect of coloured filters.

KEY WORDS

absorption
filter
opaque
reflection
transmit
translucent
transparent

White light (such as daylight) is not a single colour but a mixture of all the colours of the rainbow.

Transmission and transparency

An object that **transmits** light is either **transparent** or **translucent**. You can see through a transparent object. A translucent object transmits light but you cannot see through it. An **opaque** object does not allow light to travel through it.

How the colour of an object is determined

When light strikes an opaque object it can only be **reflected** or **absorbed**. Different surfaces reflect and absorb light of different wavelengths. The wavelengths that are reflected are the ones you see and are the colour that an object appears to be. Wavelengths not reflected are absorbed.

For example, grass absorbs all wavelengths of light except green. Green light is reflected which is why grass looks green. Each colour has its own narrow band of frequency and wavelength.

- When all wavelengths are reflected equally the object appears white.

- When all wavelengths are absorbed the object appears black.

1 What colour light is reflected by a black object?

2 What colour light is absorbed by a yellow object?

3 Explain how the colour of an object is determined.

DID YOU KNOW?

The colours in white light can be separated using a triangular prism.

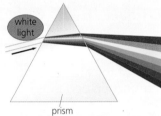

Figure 6.52 A prism splitting white light

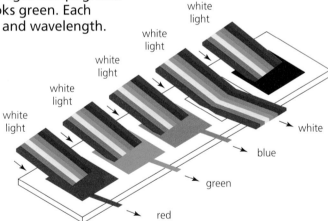

Figure 6.53 Absorption and reflection of white light by different coloured surfaces

The effect of coloured filters

Colour filters work by absorbing certain wavelengths (colours) and transmitting other wavelengths. Figure 6.54 shows the absorption effects of two different filters. If you pass white light through a red **filter**, then red light comes out the other side. This is because the red filter only allows red light through. The other colours (wavelengths) of the spectrum are absorbed. Similarly, a yellow filter only allows yellow light through.

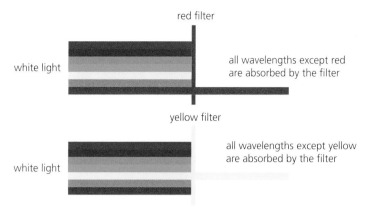

Figure 6.54 The absorption of colours in white light by a red filter and a yellow filter

The colour of an object can also change in different coloured light. The object will usually appear either the colour of the filter or black. For example, a white object reflects all wavelengths. In blue light, a white object still reflects all the wavelengths. But as there is only blue light, it looks blue.

A red object absorbs all wavelengths except red. In blue light, it absorbs all the light falling on it, so it looks black.

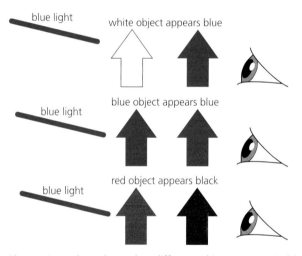

Figure 6.55 The colours that different objects appear in blue light

4 What colours of light are absorbed by a blue filter?

5 State what colour each of the following would appear through a red filter.

a a red car   c a green leaf

b a yellow flower   d a blue shirt

6 State what colour each of the following objects would appear in green light.

a a red scarf   c a black hat

b a white blouse   d a green ball

7 Suggest what happens when light hits a grey object.

Lenses

Learning objectives:

- understand what a lens does
- draw ray diagrams to show the formation of images by lenses
- describe the difference between a real and a virtual image.

KEY WORDS

concave lens
convex lens
focal length
principal axis
principal focus
real image
virtual image

The magnifying properties of natural lenses have been used for thousands of years. The Roman Emperor Nero is believed to have watched gladiators fight using an emerald lens.

DID YOU KNOW?

Lenses can be used to focus other types of electromagnetic radiation but they are most commonly used to focus light.

Convex and concave lenses

A **convex**, or converging, lens is narrow at the outside and bulges in the middle. A **concave**, or diverging, lens is narrowest in the middle and widest at the outside.

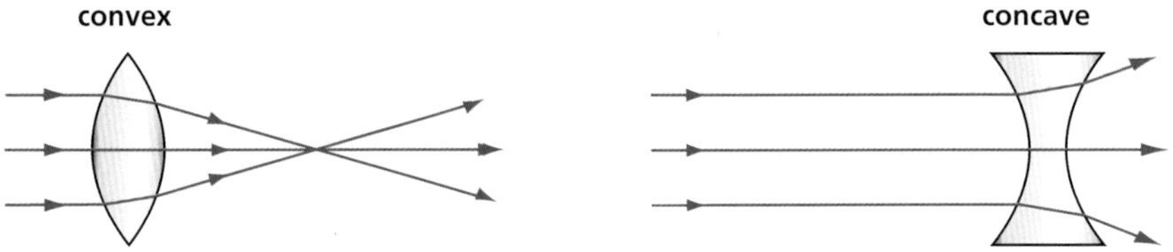

Figure 6.56 Convex and concave lenses

The symbols for convex and concave lenses are shown in Figure 6.57.

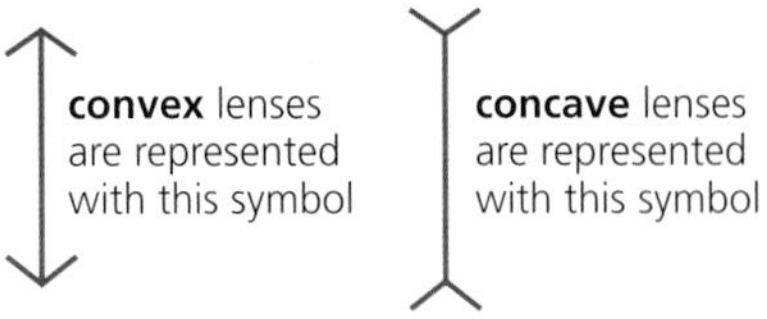

convex lenses are represented with this symbol

concave lenses are represented with this symbol

Figure 6.57 The symbols for convex and concave lenses

Refraction by a convex lens

A lens refracts light rays as they enter, then leave the lens. Figure 6.58 shows two light rays refracted towards the normal when they enter a convex lens, and away from the normal when they leave the lens. A convex lens makes rays move towards each other (converge).

The horizontal line going through the centre of a lens is called the **principal axis**. A light ray entering a lens along the principal axis is along the normal line and so will pass straight through. Light rays travelling parallel to the principal axis of a convex lens are refracted to pass through the **principal focus**, F (Figure 6.59). The light rays are focused at that point. A screen placed at the principal focus would show an image. We say the image is a **real image**, because it can be seen on a screen or other surface.

The **focal length** is the distance from the centre of the lens to the principal focus.

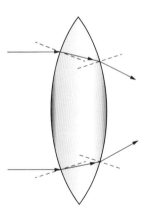

Figure 6.58 Refraction by a convex lens

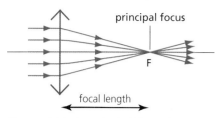

Figure 6.59 The principal focus for a convex lens

1 a What is the principal focus of a lens?

b Is it possible for rays to be focused onto a point that is not the principal focus?

2 What is the name of the straight line which goes through the centre of the lens and through the principal focus?

3 Explain how you would change the shape of a convex lens to reduce the focal length.

KEY INFORMATION

Since light can pass through a lens in either direction there are two principal foci, one on either side of the lens. These are at equal distances from the lens.

Refraction by a concave lens

A concave lens makes rays move away from each other (diverge).

Light rays travelling parallel to the principal axis of a concave lens are refracted by the lens so that they appear to have come from the principal focus (Figure 6.60). We trace back each light ray with a dashed line, because no light rays actually follow these paths.

KEY INFORMATION

The focal length is found by focusing a distant object on a piece of paper. The light rays arriving at a lens from a distant source will be almost parallel.

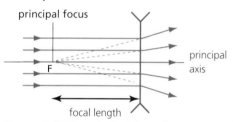

Figure 6.60 The principal focus for a concave lens

A screen placed at the principal focus of a concave lens would not show an image. We say the image is a **virtual image** because no light actually reaches it. The virtual image is at the point the light appears to have come from.

4 Explain, using the idea of refraction, why light diverges after passing through a concave lens.

5 Explain why rays from a point on a distant object are almost parallel when they reach the lens.

6 Explain why a ray that goes through the centre of a thin lens does not change direction when it passes through the lens.

Images and magnification

Learning objectives:

- draw ray diagrams to show the formation of real and virtual images by lenses
- calculate the magnification of an image.

Lenses are used in magnifying glasses, spectacles, cameras, microscopes and telescopes.

Forming a real image

To draw a **ray diagram** for a lens (Figure 6.61), draw the paths of two known rays:

- Draw a ray from the object parallel to the principal axis which refracts through the principal focus (ray 1).
- Draw a ray from the object through the centre of the lens. This ray does not change direction (ray 2).

These two rays meet at the position of the image of the object. The image is **real** (it can be projected on to a screen), inverted (upside down) and smaller.

It is helpful to draw a third ray to confirm the position of the image. Draw a ray from the object through the principal focus in front of the lens to the lens and then parallel to the principal axis.

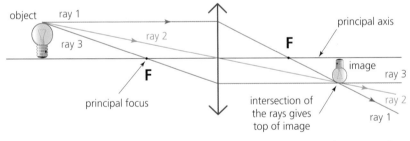

Figure 6.61 Drawing a ray diagram for a convex lens. The size and position of the image can be changed by moving the object closer to or further away from the lens

Figure 6.62 shows the ray diagram for the image formed when an object is placed between the principal focus, F, and a point at twice the focal length of the lens, 2F. The image is real, inverted and magnified (larger than the object).

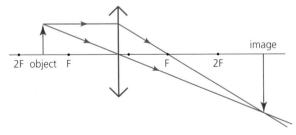

Figure 6.62 Ray diagram for an object between F and 2F

1. Draw a ray diagram for an object placed at twice the focal length of a convex lens and describe the image.

2. Draw a ray diagram for an object located at the focal point and describe the image.

3. In a camera, a convex lens focuses light rays to produce an image on the film or sensor. Draw a ray diagram to show how a camera lens produces a real image of a distant object.

Forming a virtual image

When an object is placed between the principal focus and a convex lens a **virtual image** is produced. The image is the right way up and magnified (Figure 6.63). The dashed lines show virtual rays used to construct the position of the image.

A concave lens always forms a virtual image that is the right way up and smaller than the object. Figure 6.64 shows how to draw this ray diagram.

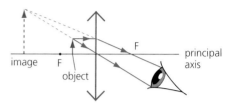

Figure 6.63 Drawing a ray diagram for the formation of a virtual image by a convex lens when the object is between the lens and its focal point

4. Draw a ray diagram for an object placed at a distance 2F from a concave lens and describe the image.

Magnification

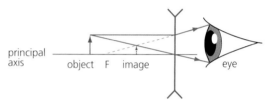

Figure 6.64 The ray diagram for a concave lens producing a virtual image

If a lens is used to produce an image that is larger than the object, the **magnification** is how many times bigger the image is than the object. The magnification can be calculated using the equation:

$$\text{magnification} = \frac{\text{image height}}{\text{object height}}$$

Example: If an object was 2 mm long and a hand lens was used to make it look 1 cm long, the magnification would be:

$$\frac{\text{image height}}{\text{object height}} = \frac{10\,\text{mm}}{2\,\text{mm}} = 5$$

KEY INFORMATION

Magnification is a ratio and so has no units. Image and object height must both be measured in the same units, e.g. mm or cm.

5. An insect wing 1.6 mm long is viewed with a magnifying glass, and appears to be 1.2 cm long. What is the magnification produced by the lens?

6. A lens is used to focus an image of person 1.76 m tall onto a white wall. The image on the wall is 1.9 cm tall. Calculate the magnification produced by the lens.

KEY INFORMATION

If the image is smaller than the object the magnification is <1.

7. An object 1 cm high was placed 3 cm in front of a convex lens. The focal length of the lens is 2 cm.

Draw a ray diagram to determine the magnification.

Emission and absorption of infrared radiation

Learning objectives:

- realise that all bodies emit and absorb infrared radiation
- compare emission and absorption of radiation from different surfaces
- define a perfect black body
- explain that the intensity and distribution of wavelengths of any emission depend on the temperature of the body.

Some of the coldest objects in the Universe, such as the dense clouds of gas and dust in which new stars are formed, have temperatures of just a few degrees above absolute zero. They emit most of their radiation at wavelengths of around a millimetre. These wavelengths are invisible to the human eye, but can be detected by radio telescopes.

Infrared radiation

All objects, whatever their temperature, both **emit** (give out) and **absorb** (take in) **infrared radiation**. The hotter an object, the more infrared radiation it emits in a given time (the greater its intensity). This is why objects at different temperatures appear at different brightnesses on a thermal imaging camera.

When infrared radiation strikes an object, the radiation may be absorbed by it, transmitted (pass through it) or reflected back.

1. Which object in the photo in Figure 6.65 is emitting the most infrared radiation? Why is this?

2. Why are other objects in the photo also emitting infrared radiation?

Good absorbers and emitters

The surfaces of different objects at the same temperature affect how well each surface absorbs or emits infrared radiation. Figure 6.66 shows an experiment you may have done to compare the emission of infrared radiation from different sides of a container of hot water. The sides will all be at the same temperature as the water. However, if you hold your hand the same distance from each side you would find that the dull (matt) black surface feels warmer than the shiny white surface.

An infrared sensor held a fixed distance from each side can measure the rates at which energy is emitted (as infrared radiation) from each surface. Each side has the same area. The dull black surface should give the highest reading. This shows

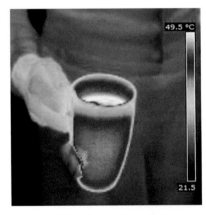

Figure 6.65 A thermal imaging camera detects infrared radiation. The camera converts the intensity of infra-red radiation to different colours

COMMON MISCONCEPTION

Remember that infrared radiation is invisible. You *see* red light but you *feel* infrared radiation.

that the matt black surface is the best emitter of infrared radiation. It emits radiation at a faster rate. White or silvery surfaces are poor emitters.

3 Which surface in Figure 6.66 should give the lowest reading on the infrared sensor?

A good emitter of electromagnetic radiation is also a good absorber of electromagnetic radiation. White, shiny surfaces are poor at absorbing and emitting infrared radiation.

4 Solar hot water panels heat water for the home by absorbing infrared radiation from the Sun. Water is slowly pumped through copper pipes inside the solar panel. Explain why it is useful to paint the shiny copper pipes black.

A perfect black body

Black surfaces are good absorbers of radiation. Imagine an object which absorbed all radiation that falls onto it. As all the radiation is absorbed, none is reflected or transmitted. This is called a perfect **black body**. It appears perfectly black.

Since a good absorber is also a good emitter, a perfect black body would be the best possible emitter.

At a given temperature, a perfect black body radiates at all wavelengths. The intensity and peak wavelength of the radiation emitted depend only on the object's temperature. The hotter the object, the more radiation it radiates in a given time. This means the intensity of every wavelength emitted increases with temperature – the curves in Figure 6.67 all get higher as temperature increases. However the relative brightness of the different colours emitted also changes as the temperature increases – the curves in Figure 6.67 shift to shorter wavelengths as temperature increases.

This change in the intensity of wavelengths emitted explains why a heated metal object appears first red, then yellow, then white.

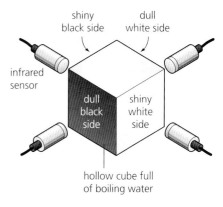

Figure 6.66 An experiment to compare emitters of infrared radiation

DID YOU KNOW?

In hot countries, houses are often painted white to keep them cool inside.

5 Describe what a black object would look like if it was heated up to a temperature much higher than 7200 °C.

6 What would a white object look like if it was heated up to the same temperature as the object in question 5?

KEY INFORMATION

All objects emit radiation, and the intensity and the wavelength distribution always depend on the temperature. But for a perfect black body these depend *only* on the temperature.

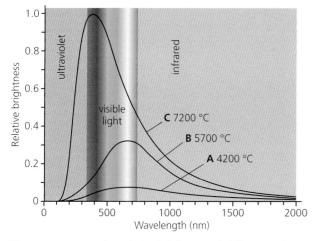

Figure 6.67 How the relative brightness of different colours emitted changes with temperature of the object

Temperature of the Earth

Learning objectives:

- describe how the atmosphere absorbs radiation in a way that varies with wavelength
- list the factors affecting the temperature of the Earth
- explain how the temperature of an object is related to the radiation absorbed and radiation emitted.

Although the temperature varies over the Earth's surface, the mean surface temperature on Earth is approximately 14 °C. The hottest surface temperature ever recorded on Earth was 93.9 °C, in Furnace Creek, Death Valley, California.

Radiation balance

Objects that are hotter than their surroundings emit **infrared radiation** at a greater rate than they **absorb** it, and so transfer energy to their surroundings. Objects that are cooler than their surroundings **emit** some infrared radiation, but they absorb infrared radiation at a greater rate than they emit it.

When the rates at which radiation is absorbed and emitted are the same, the temperature of an object does not change. This is called thermal equilibrium.

1 **When an object is absorbing infrared radiation faster than it emits it what can you say about its temperature?**

2 **Explain why white objects take longer to reach thermal equilibrium with their surroundings than black objects do.**

Temperature of the Earth

The Earth absorbs radiation from the Sun and it also emits infrared radiation because its temperature is higher than that of its surroundings.

Most of the energy the Sun radiates is in the visible part of the electromagnetic spectrum. Visible light can pass through the Earth's atmosphere – this is how we see the Sun and stars. So most solar radiation is transmitted through the atmosphere before being absorbed by the surface of the Earth. Some of the incoming radiation is also absorbed or reflected by the atmosphere.

The Earth then re-radiates the energy it absorbed, as infrared radiation. Some of

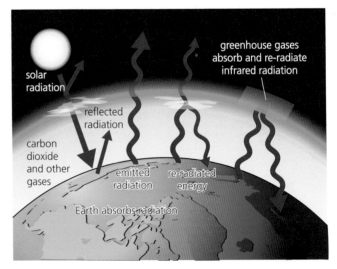

Figure 6.68 For the mean temperature on Earth to stay constant there must be a balance between the rates of absorption, emission and reflection of radiation

this outgoing radiation is absorbed by greenhouse gases such as carbon dioxide, methane and nitrogen oxides, which occur naturally in the atmosphere as well as being produced by human activities.

The temperature of the Earth depends on the rate of absorption of radiation from the Sun, the rate of emission of radiation and the rate of reflection of radiation into space.

3 Why does the Earth emit infrared radiation?

4 State what factors the mean temperature of the Earth depends on.

Changing the Earth's radiation balance

The temperature of the Earth can be altered by changing the rates of absorption, emission and reflection of radiation. The rates can be affected by natural processes and by the activities of humans.

Changing the types and levels of greenhouse gases changes how much infrared radiation is absorbed in the atmosphere. If less of the outgoing infrared radiation (emitted from the Earth's surface) is absorbed, more 'escapes' from the Earth so the surface temperature would decrease.

The amount of cloud cover can also affect the rate of **reflection** of solar radiation. When there is more cloud cover, more solar radiation is reflected away from the Earth so less is absorbed by the Earth's surface or atmosphere.

The activities of humans are having a much greater effect than natural processes. Increased levels of greenhouse gases in the atmosphere are due to factors such as burning fossil fuels, deforestation, the increased use of chemical fertilisers and the increased use of farm animals.

5 a Suggest some natural effects on the rates of absorption, reflection and emission of radiation on Earth.

b Suggest some activities of humans that affect the rates.

6 Suggest how human activities affect the rate of emission of radiation from the Earth.

7 An eruption in Iceland in 1783 caused there to be many particles in the atmosphere over a large area for several months. Suggest what effect this would have had on the temperature of parts of the Earth.

8 Give reasons why it is difficult to prove that increased levels of carbon dioxide are raising global temperatures.

DID YOU KNOW?

The level of carbon dioxide in the atmosphere is measured at the Mauna Loa observatory in Hawaii. The annual mean amount of carbon dioxide has increased from about 320 parts per million in 1960 to just over 400 parts per million in 2015.

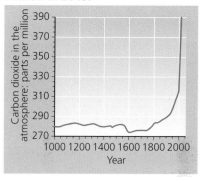

Figure 6.69 Levels of carbon dioxide in the atmosphere

MATHS SKILLS

Using and rearranging equations

Learning objectives:

- select and apply the equations $T = \frac{1}{f}$ and $v = f\lambda$
- substitute numerical values into equations using appropriate units
- change the subject of an equation.

The names of units are often taken from the name of the scientist who first worked in that particular field of physics. The unit of frequency is the hertz (Hz) after the German physicist Heinrich Hertz.

Period and frequency

Remember, frequency, f, is the number of waves passing a point each second. The unit of frequency is the hertz (Hz), which means cycles per second. The time period, T, is measured in seconds (s).

Period and frequency are linked by the equation:

$$\text{period, } T = \frac{1}{\text{frequency, } f}$$

Example: A wave has a frequency of 5 Hz. What is its period?

Substituting 5 Hz into the equation $T = \frac{1}{f}$ gives $T = \frac{1}{5}$ s. When there are 5 cycles of a wave in 1 second, then one cycle, or the period T, is $\frac{1}{5\,\text{Hz}}$ or 0.2 s.

① **Work out the period of a wave when the frequency is:**

    a 100 Hz   b 1000 Hz   c 15 000 Hz.

Example: Calculate the frequency when the period is 4 seconds. Use the equation $T = \frac{1}{f}$.

Rearrange it to make f the **subject of the equation.**

Multiply both sides by f: $Tf = 1$

Divide both sides by T: $f = \frac{1}{T}$

Substitute $T = 4$ s into the rearranged equation:

$$f = \frac{1}{4\,s} = 0.25 \text{ Hz}$$

② **Work out the frequency of a wave with period:**

    a 5 s   b 10 s   c 150 s.

KEY INFORMATION

You do not need to remember the equations on this page as they will be on the equation sheet. But you need to know when and how to use each equation. You also need to be able to rearrange an equation. Rearranging an equation means making another variable the subject of the equation. The subject of the equation is on its own, usually on the left-hand side.

Speed, frequency and wavelength

You can calculate the speed at which a wave moves using the equation:

$$speed = \frac{distance}{time}$$

The unit of distance is the metre (m). The unit of time is the second (s). So the unit of wave speed is metres per second (m/s).

Wavelength is the distance from a point on one wave to the equivalent point on the adjacent wave, such as between two adjacent crests. The symbol for wavelength is λ and the unit is the metre (m).

The wave equation links wave speed, frequency and wavelength:

wave speed = frequency × wavelength

$v = f\lambda$

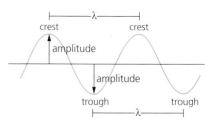

Figure 6.70 Wavelength of a transverse wave

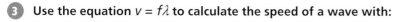

3 Use the equation $v = f\lambda$ to calculate the speed of a wave with:

  a frequency 100 Hz and wavelength 2 m

  b frequency 100 Hz and wavelength 2 cm

  c frequency 100 Hz and wavelength 2 mm.

4 Rearrange the wave equation to make frequency the subject.

5 Calculate the frequency of a wave with:

  a wavelength 0.5 m and speed 25 m/s

  b wavelength 0.05 m and speed 250 m/s

  c wavelength 0.005 m and speed 2500 m/s.

6 a Rearrange the wave equation to calculate the wavelength given the speed and the frequency.

  b Calculate the wavelength of a sound wave in air with frequency 500 Hz and speed 330 m/s.

> **MATHS**
>
> When you rearrange an equation, always do the same operation (such as multiplication or division) to both sides. For example, if you divide one side of the equation by the variable λ, you must divide both sides of the equation by λ.

Changes in velocity, frequency and wavelength

Sound travels at different speeds through different media. The denser the material, the closer the molecules and the quicker the vibrations will travel through that substance.

If speed increases when a sound wave is transmitted from one medium to another, what effect does that have on the wave's frequency and wavelength?

If speed increases then either frequency or wavelength must also increase. This is because in the wave equation, wave speed is **proportional** to frequency and wavelength.

7 Red light has a wavelength of 6.5×10^{-7} m and moves a distance of 3.0 m in 1.0×10^{-8} s. Calculate the frequency of red light.

Material	Speed of sound (m/s)
air	330
water	1500
wood	3300
titanium	6100

8 A sound wave has a wavelength of 25 cm in air.

Use the data in the table to calculate its wavelength in titanium. Assume that the frequency of the wave does not change.

Check your progress

You should be able to:

Describe the amplitude, wavelength, frequency and period of a wave →	Use the wave equation $v = f\lambda$ to calculate wave speed →	Rearrange and apply the wave equation
Realise that waves can be transverse or longitudinal →	Give examples of longitudinal and transverse waves →	Explain the difference between transverse and longitudinal waves
Describe how sound waves travel through air or solids →	Describe how to measure the speed of sound waves in air →	Explain how to calculate the depth of water using echo sounding
Understand that waves transfer energy or information →	Give examples of energy transfer by waves (including electromagnetic waves) →	Describe evidence that, for e.g. ripples on a water surface, it is the wave and not the water itself that travels
Understand that waves can be absorbed, transmitted or reflected at a surface →	Describe examples of reflection, transmission and absorption of waves (including electromagnetic waves) at material interfaces →	Describe how different substances may absorb, transmit, refract or reflect electromagnetic waves in ways that vary with wavelength
Draw a labelled ray diagram to illustrate reflection of a wave at a boundary →	Construct ray diagrams to illustrate refraction at a boundary →	Use wavefront diagrams to explain refraction in terms of a change in wave velocity
Describe the range of normal human hearing. Define the term ultrasound →	Describe how ultrasound waves can be used for medical and industrial imaging →	Explain how P and S waves can be used to deduce information about the structure of the Earth
Name the main groupings of the electromagnetic spectrum →	Compare the groupings of the electromagnetic spectrum in terms of wavelength and frequency →	Describe how radio waves are produced
Describe the hazardous effects of gamma rays, X-rays and ultraviolet radiation →	Explain the risks associated with the use of ionising and ultraviolet radiation →	Evaluate the risks and consequences of exposure to radiation
Give examples of the uses of the main groupings of the electromagnetic spectrum →	Describe examples of energy transfer by electromagnetic waves →	Explain why each type of electromagnetic wave is suitable for the application
State that each colour in the visible spectrum has its own narrow band of wavelength →	Describe that colour filters absorb certain wavelengths and transmit other wavelengths →	Explain that the colour of an opaque object depends on which wavelengths are more strongly reflected
State that in a convex lens parallel rays of light are brought to a focus at the principal focus →	Use ray diagrams to determine the nature of the image formed by a lens →	Use ray diagrams to determine the position and magnification of images
State that the hotter the body the more radiation it emits in a given time →	Explain that a perfect black body absorbs all the radiation incident on it, and does not reflect or transmit any radiation →	Explain how the temperature of a body is related to the balance between incoming radiation absorbed and radiation emitted

Worked example

The table below shows the electromagnetic spectrum.

A	microwave	infrared	visible light	B	X-rays	gamma rays

1 **State the names of the waves labelled A and B.**

A = radio waves   B = ultraviolet waves

Both answers are correct. Use a mnemonic to remember the correct order.

2 **X-rays are dangerous to humans. Explain how they can also be used in medical therapy without lasting harm.**

We can use them for X-rays to see our bones and for treating cancer by killing the cancer cells.

This answer is a good start but is incomplete. It doesn't give a full explanation. The answer should also say that we can use them by controlling the exposure dose.

3 **Complete the ray diagram to show how the image is formed.**

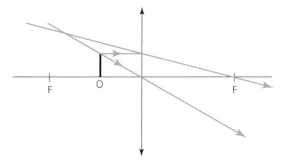

The positions of the rays have been drawn correctly, but the virtual rays have not been drawn with dashed lines. Also the image has not been drawn in. You should always draw the image when you have completed the ray diagram.

4 **Explain why we see a red apple as red.**

White light is a mixture of many wavelengths. We see the red apple because the red wavelengths are reflected off the apple.

This is a good answer. Just one thing has been missed out: red wavelengths are reflected but what about the other wavelengths? They are absorbed.

5 **Complete the wave front diagram to show the refraction of light from air to water.**

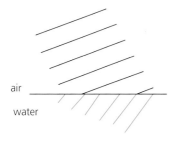

air

water

This correctly shows the wavelength being less but as the waves are entering a denser medium they will be refracted towards the normal. The wavefronts change direction but they still match up (see Fig 6.37).

End of chapter questions

Getting started

1. Which of the following is a longitudinal wave? `1 Mark`
   a sound wave   b water wave   c light wave   d radio wave

2. Which type of lens brings rays together at the principal focus? `1 Mark`

3. Which is a correct unit of frequency? `1 Mark`
   a metres   b watts   c hertz   d metres per second

4. Explain what is meant by the **amplitude** of a wave. `2 Marks`

5. Which electromagnetic wave has the shortest wavelength? `1 Marks`

6. A water wave in a ripple tank has a frequency of 4 Hz. Calculate the time period (T) of the wave using the equation $T = 1/f$. `1 Mark`

7. A microwave has a wavelength of 1 cm. What type of wave, other than a microwave, might have a wavelength of 1 m? `1 Mark`

8. A ray of light is shone onto a dull, white object but no reflected ray is seen. Explain what type of reflection this is and what happens to the light. `2 Marks`

Going further

9. Name the parts of the electromagnetic spectrum labelled A and B in the diagram. `1 Mark`

   | radio waves | microwave | A | visible light | ultraviolet | B | gamma rays |

10. Name the two types of electromagnetic radiation that are used to cook food. `1 Mark`

11. Describe two practical applications for:

    a microwaves

    b gamma rays. `2 Marks`

12. Calculate the magnification of a lens when the image height is 2.5 mm and the object height is 7.5 mm. `2 Marks`

13. What is the speed of a water wave if it has a wavelength of 8 cm and a frequency of 2 Hz? `2 Marks`

14. In hot countries, houses are often painted white. Explain why this is done. `2 Marks`

More challenging

15. Explain why you can't hear a dog whistle with a frequency of 25 000 Hz. `1 Mark`

16. State what colours of light make a blue T-shirt look black. `1 Mark`

17. Explain why houses would benefit in both hot and cold countries if they had a white roof. `2 Marks`

18   a   Construct a ray diagram for the image of an object in a convex lens that is nearer to the lens than one focal length.   `2 Marks`

     b   Describe the image produced by the lens.   `2 Marks`

     c   Long-sighted people struggle to see things that are close to their eyes but can see things clearly if they are further away. Explain how your ray diagram shows why convex lenses are used in glasses for long-sighted people.   `2 Marks`

Most demanding

19   A survey ship sailing in a straight line at a speed of 4 m/s sends out a sound pulse from its echo sounder every 10 s. The time between each pulse being sent and its reflection are as follows: 0.25 s, 0.30 s, 0.35 s, 0.40 s, 0.40 s, 0.40 s, 0.40 s, 0.05 s, 0,40 s, 0.35 s, 0.30 s, 0.25 s, 0.25 s and 0.25 s.

   Use this data to describe the sea bed the ship is sailing over and suggest a reason for the apparent anomalous result.   `2 Marks`

20   An ultrasound scanner was used to measure the size of a kidney stone. Use the following data to determine the thickness of the stone.   `2 Marks`

   Speed of ultrasound in the stone = 4.00 km/s

   Time between emitted ultrasound pulse and reflection from the front of the stone = 0.0500 ms

   Time between emitted ultrasound pulse and reflection from the back of the stone = 0.0600 ms

21   The diagram below gives information about the effect of radiation on the temperature of the Earth. Explain the conclusions you can draw from this, including how greenhouse gases might contribute to global warming.   `6 Marks`

solar radiation and IR emission

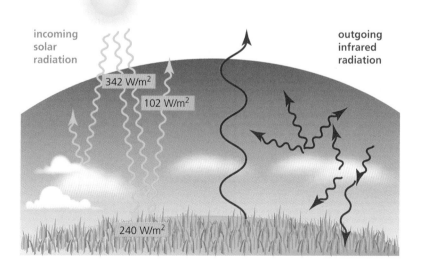

`Total: 40 Marks`

ELECTROMAGNETISM

IDEAS YOU HAVE MET BEFORE:

PERMANENT MAGNETS

- Permanent magnets are made from nickel, cobalt and most steels.
- Iron is a magnetic material but loses its magnetism when the magnetising force is removed.
- Like poles repel and unlike poles attract.
- Repulsion is the only test for a magnet.

EARTH'S MAGNETISM

EARTH'S MAGNETIC FIELD

- The Earth's magnetic field is very similar to the magnetic field due to a bar magnet.
- A compass uses the Earth's magnetic field to indicate magnetic north.
- The Earth behaves as if there is a gigantic bar magnet inside it.

MAGNETIC EFFECT OF A CURRENT

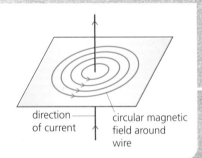

direction of current

circular magnetic field around wire

- The wires carrying an electric current produce a magnetic field.
- The magnetic field made by the current in a wire is a circular shape around the wire.
- The magnetic field is increased if more turns are wound on the coil and/or the current in the wire is increased.

ELECTROMAGNETS

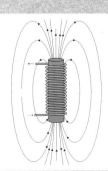

- A coil of wire with an iron rod in it makes an electromagnet.
- The iron rod becomes magnetised when a current flows in the coil.
- The iron rod becomes strongly magnetised but loses its magnetism as soon as the current is switched off.

IN THIS CHAPTER YOU WILL FIND OUT ABOUT:

WHAT IS A MOTOR AND HOW DOES IT WORK?

- Describe some uses of motors.
- Explain how a motor works.
- Explain how a commutator is used.

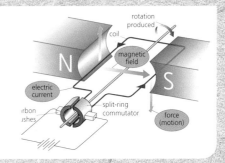

HOW CAN A MAGNETIC FIELD BE USED TO PRODUCE AN ELECTRIC CURRENT?

- Dc generators are similar in construction to dc motors.
- A coil is rotated in a magnetic field. This induces an alternating current in the coil.
- Direct current like that from a battery is obtained by using a commutator.

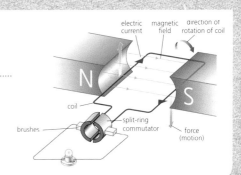

WHAT IS A TRANSFORMER?

- A transformer changes the size of an ac potential difference.
- Step-up transformers increase the potential difference.
- Step-down transformers decrease the potential difference.

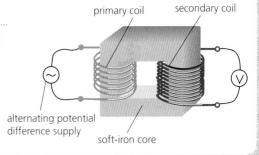

HOW CAN WE USE TRANSFORMERS TO SUPPLY ENERGY EFFICIENTLY?

- Using a transformer to increase the potential difference reduces the current.
- A smaller current in the transmission lines reduces the energy losses to the surroundings.
- This means a greater proportion of the energy reaches the end-user.

Magnetism and magnetic forces

Learning objectives:

- explain what is meant by the poles of a magnet
- plot the magnetic field around a bar magnet
- describe magnetic materials and induced magnetism.

KEY WORDS

attract
induced magnet
magnetic field
permanent magnet
poles
repel

A maglev (magnetic levitation) train has no wheels. It uses the fact that like poles repel so that it hovers above its track. The maglev train in Shanghai can carry passengers 30 km to the airport in 7 minutes and 20 seconds (Figure 7.1).

Figure 7.1 A maglev train in Shanghai

The poles of a magnet

The **poles** of a magnet are the places where the magnetic forces are strongest. When a bar magnet is suspended by a thread, it will settle with one end pointing north. This end is called the north-seeking pole, or north pole.

When the ends of two bar magnets are brought near each other, two north poles or two south poles **repel** each other but a south pole and a north pole **attract** each other (Figure 7.2). These are examples of non-contact forces.

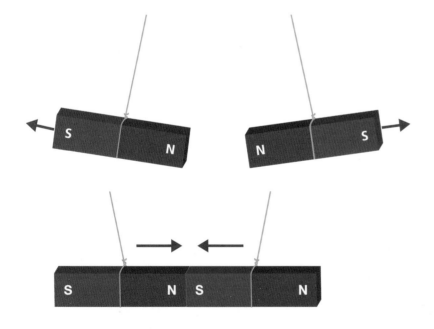

KEY INFORMATION

Magnetic field lines always start at a north pole and finish at a south pole. The lines show the direction of the magnetic field.

Figure 7.2 Like poles repel, unlike poles attract

1 **What happens when the south poles of two magnets are brought together?**

Magnetic fields

A **magnetic field** is the region around a magnet where a force acts on another magnet or on a magnetic material. The direction of a magnetic field is the direction of the force on a north pole placed at a point in the field.

You can use a plotting compass to reveal the magnetic field of a magnet. Put the magnet on a piece of paper and draw round it. Place the plotting compass near the north pole of the magnet. Draw a dot by the head of the arrow on the compass. Move the compass so that the tail of the pointer is by the dot you have just drawn. Draw another dot by the head of the arrow.

Repeat moving the compass and drawing a dot by the head of the arrow until you have gone all the way round to the south pole. Draw a smooth curve to join all your points.

Repeat for slightly different starting points near to the north pole of the magnet.

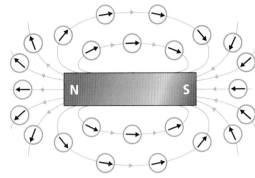

Figure 7.3 The needle of a plotting compass can be used to show the direction of the magnetic field around a bar magnet

2 Explain why you need to put the plotting compass in several different places near the north pole of the magnet when using it to plot the magnetic field.

Induced magnetism and magnetic materials

Iron, some steels, nickel and cobalt are magnetic materials. They can all be magnetised. Iron only makes a temporary magnet. It loses its magnetism as soon as the magnetising force is removed.

A permanent magnet produces its own magnetic field. An **induced magnet** is a material that becomes a magnet when it is placed in a magnetic field. When removed from the magnetic field it loses most or all of its magnetism quickly. Iron is an example of this type of magnetic material. **Permanent magnets** are made from nickel, cobalt and some types of steel.

The force between a magnet and a magnetic material is always attractive.

Steel pins or iron tacks are attracted to a magnet (Figure 7.4). Each pin or tack becomes a small magnet because of induced magnetism. If you carefully separate the magnet from the steel pins and tacks, the steel pins remain magnetised but the iron tacks lose their magnetism and are no longer attracted to each other.

You can test whether a material is a magnet by bringing another magnet close to it.

> **KEY INFORMATION**
>
> **Magnetic field lines never meet or cross.**

Figure 7.4 Magnets attract some metals such as iron and steel

3 Explain why the pins and tacks behave as described.

4 Explain what forces of attraction and repulsion can take place between the pins and tacks once they have been removed from the magnet (Figure 7.4).

5 Suggest how you know whether an object is a magnet when you bring another magnet close to it.

6 Suggest how the behaviour of a magnetic compass is related to evidence that the core of the Earth must be magnetic.

Compasses and magnetic fields

Learning objectives:

- describe the Earth's magnetic field
- describe the magnetic effect of a current
- explain the link between current and magnetic field.

Objects like steel girders in bridges are often found to be slightly magnetised by induction from the Earth's magnetism, particularly if the structure is hammered or shaken by vibrations.

Magnetic compass

A compass contains a small bar magnet. One end is a north pole and one end is a south pole. The south pole of a compass needle is attracted to a magnetic north pole. This is how we use a compass to plot the magnetic field of another magnet (Figure 7.3 in topic 7.1).

1 **Draw a labelled diagram to show how a compass needle will behave when placed near:**

   a a north pole

   b a south pole.

The Earth's magnetic field

The fact that a compass needle points north is evidence that the Earth has a magnetic field. The Earth behaves as if there is a bar magnet inside it. (Such a magnet cannot really exist because the centre of the Earth is too hot.) The odd thing is that the south pole of this imaginary magnet is in the northern hemisphere near the Earth's geographic north pole (Figure 7.5). The geographic north pole is a magnetic south pole. The end of the compass that points towards geographic north is called a north-seeking pole.

DID YOU KNOW?

....

At the Earth's centre is a solid inner core surrounded by a fluid outer core, which is hottest next to the inner core. Hot molten iron rises within the outer core, then cools and sinks. These movements create an electric current, which, combined with the rotation of the Earth, are thought to produce the Earth's magnetic field.

DID YOU KNOW?

....

The magnetic poles move around over time, and the **Earth's magnetic field** has reversed direction many times in the Earth's history. These flips are becoming more frequent. Many millions of years ago, the field reversed direction every 5 million years, but now it reverses approximately every 200 000 years.

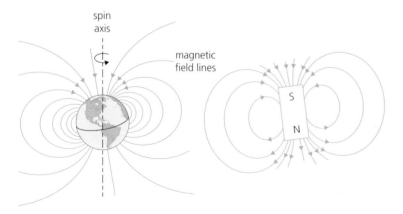

Figure 7.5 The Earth's magnetic and geographic poles do not exactly coincide

The closer the field lines are together, the higher the strength of the magnetic field. Figure 7.5 shows that the Earth's magnetic field is strongest at the poles.

2 **What is the polarity of the end of the imaginary bar magnet inside the Earth that is nearest the geographic south pole?**

3 **Explain why the north pole of a compass is often called the north-seeking pole.**

The magnetic effect of a current

When a current flows in a wire, the current creates a magnetic field around the wire. This effect can be investigated using a plotting compass placed at different points around the wire (Figure 7.6).

DID YOU KNOW?

When Hans Christian Oersted discovered that electricity produced magnetism, scientists started to look for the reverse effect. In 1831, Michael Faraday discovered how to produce electricity using magnetism.

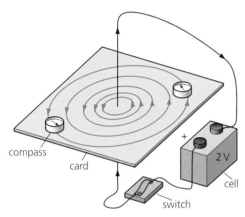

Figure 7.6 The magnetic field around a current-carrying wire is circular

The strength of the magnetic field decreases with distance from the wire. Further from the wire, the field lines would get further apart. The strength of the magnetic field also depends on the current through the wire. The higher the current, the stronger the magnetic field.

If the current direction is reversed, the direction of the magnetic field is reversed. The direction of the magnetic field is given by the right-hand grip rule (Figure 7.7).

To work out the direction of the magnetic field, grip the wire in your right hand so that your thumb points in the direction of the conventional flow of current (from positive to negative). Then your fingers point around the wire in the direction of the magnetic field.

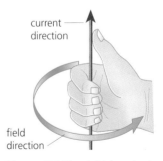

Figure 7.7 The right-hand grip rule

4 **A wire goes into the plane of a page. Draw a diagram to show the magnetic field and its direction around the wire.**

5 **Copy Figure 7.6 and sketch a diagram to show how the magnetic field pattern would change if the current in the wire was increased.**

6 **Explain why a compass placed towards the edge of the card in Figure 7.6 might not point along the field line as expected.**

The magnetic effect of a solenoid

Learning objectives:

- draw the magnetic field around a conducting wire and a solenoid
- describe the force on a wire in a magnetic field
- apply the left-hand rule to work out the direction of a magnetic field, a current or a force around a wire.

KEY WORDS

Fleming's left-hand rule
motor effect
solenoid

Electricity and magnetism are very closely connected. By combining the two, we can create much more powerful magnets or electromagnets.

Magnetic field of a loop of wire

You have seen how a current flowing through a wire creates a magnetic field around it. The magnetic field becomes more complicated when the wire becomes a loop. The field lines are squashed together in the middle of the loop.

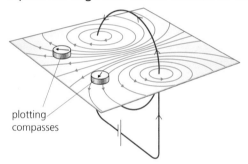

Figure 7.8 The magnetic field in the centre of a loop is stronger than outside the loop

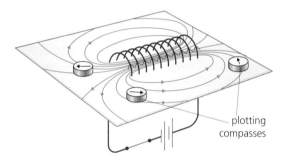

Figure 7.9 The magnetic field around a solenoid is similar to that of a bar magnet

1 The direction of the current in the wire in Figure 7.8 is reversed. What is the direction of the magnetic field now?

The magnetic field in a solenoid

A **solenoid** is a long straight coil of wire. When a current is passed through a solenoid the magnetic field of all the coils combines to produce a magnetic field like that of a bar magnet.

The polarity of the solenoid can be worked out by looking at the end of the solenoid and seeing which way the current passes (Figure 7.10).

2 Sadie looks at the end of a solenoid and notes that the current is in a clockwise direction. What is the magnetic pole at that end of the solenoid?

3 Suggest two things that you can do to make the magnetic field stronger.

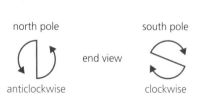

Figure 7.10 North and south poles of a coil of wire

The kicking wire

When a straight wire with a current passing through it is placed between the poles of a magnet the two magnetic fields combine, making the resultant magnetic field stronger in one area and weaker in another area. This produces a force on the wire which makes the wire move (Figure 7.11). This is called the **motor effect**. The direction of the force and the motion is perpendicular to both the magnetic field and the current.

Fleming's left-hand rule (Figure 7.12) allows us to work out the relative directions of the current, magnetic field and motion of the wire.

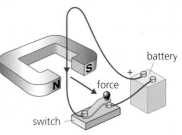

Figure 7.11

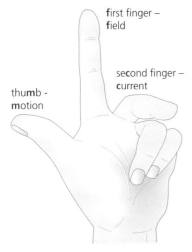

Figure 7.12 Fleming's left-hand rule

> **REMEMBER!**
>
> The right-hand grip rule and Fleming's left-hand rule use different hands. Be careful not to muddle them up.

> **KEY INFORMATION**
>
> Conventional current is from positive to negative terminal of a battery. This is the *opposite* direction to the direction of electron flow.

Hold your thumb and first two fingers of your left hand at right angles to each other.

Point your first finger in the direction of the magnetic field (from north to south).

Point your second finger in the direction of the current. This is what is known as conventional current – flowing from positive to negative.

Your thumb then shows the direction of motion of the wire.

You can increase the force on the wire by increasing:

- the size of the magnetic field
- the current passing through the wire
- the length of the wire that is in the magnetic field.

4 A wire is placed between the poles of a magnet. When the current is switched on the wire moves to the left.

  a Which way does the wire move when the current direction is reversed?

  b Which way does the wire move when the poles of the magnet are reversed?

5 Explain how you could decrease the force on the wire.

6 Which way will the wire move when the switch is pressed in Figure 7.11?

Electromagnets in action

Learning objectives:

- describe simple uses of electromagnets
- explain how an electric bell works
- interpret diagrams of other devices that use electromagnets to explain how they work.

The starter motor in a car needs a very large current (100 A or more). A relay is used which contains an electromagnet. This means that the large current does not flow in the same circuit that contains the ignition key.

Using simple electromagnets

In a simple **electromagnet** (Figure 7.13) a coil of copper wire (solenoid) is wound round an iron core. This magnetises the core, creating an electromagnet about 1000 times stronger than the coil by itself. An electromagnet can be switched on and off because the iron core is a a temporary magnet. It loses its induced magnetism when the current is switched off, because there is no longer a magnetic field caused by the current in the wire.

DID YOU KNOW?

An electromagnet used in a scrapyard can lift up to 15 tonnes of metal.

1. Why is the core of an electromagnet made of iron?

2. Explain what would happen if the core was made from steel.

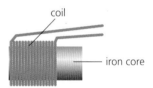

Figure 7.13 An electromagnet can be switched on and off

The electric bell

Electric bells contain an electromagnet (Figure 7.14). Pushing the switch (bell-push) completes a circuit which magnetises the electromagnet and makes the hammer hit the gong.

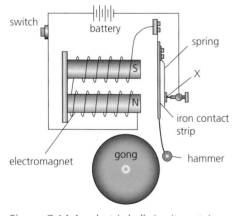

Figure 7.14 An electric bell circuit contains an electromagnet

KEY INFORMATION

The strength of an electromagnet is increased by adding an iron core. The strength also depends on the current through the coil, the material in the core, and the number of turns on the coil.

When the bell-push completes the circuit a current passes through the electromagnet. The iron strip is attracted to the electromagnet and the hammer hits the gong. This movement breaks the circuit at X, so the current stops flowing and switches off the electromagnet. The spring pulls the strip back, so that contact is made and the sequence begins again.

This is called a make-and-break circuit.

3 Suggest why Figure 7.14 is called a make-and-break circuit.

4 Explain what makes the hammer strike the gong.

5 Explain why the hammer moves back again after the gong has been struck.

The relay

You do not need to remember how a relay works, but you should be able to interpret diagrams of electromagnetic devices in order to explain how they work.

In Figure 7.15, the magnetic strip on the card operates a switch so that a current passes through a solenoid and opens the door.

A relay is a magnetic switch. It uses a small current to switch on a much larger current.

The starter motor in a car needs a very large current (100 A or more). A relay is used so that this large current is switched on by a much smaller current in the dashboard switch.

Figure 7.15 Many hotels issue cards with magnetic strips to guests as keys to their rooms

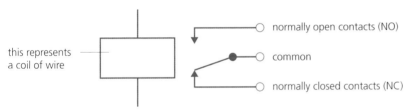

Figure 7.16 A relay is used to switch on a large current using a much smaller current

6 Suggest why using a relay for the starter motor in a car is much safer than putting the switch in the same circuit as the starter motor.

7 Suggest how you could adapt the electric bell in Figure 7.14 to make a relay. Assume the hammer and the gong are made from metal.

Calculating the force on a conductor

Learning objectives:

- explain the meaning of magnetic flux density, *B*
- calculate the force on a current-carrying conductor in a magnetic field.

A conducting wire in a magnetic field experiences a force – the motor effect. The application of this effect is the basis for all electric motors.

HIGHER TIER ONLY

Magnetic flux density, B

The spacing of the magnetic field lines in Figure 7.17 tells us how strong the magnetic field is. Where the lines are closer together, the field is stronger. The magnetic field inside the coil is much stronger than the magnetic field outside the coil.

The strength of a magnetic field is called the **magnetic flux density**, denoted by *B* and measured in **tesla (T)**.

1 **Explain how the strength of the magnetic field varies inside the solenoid.**

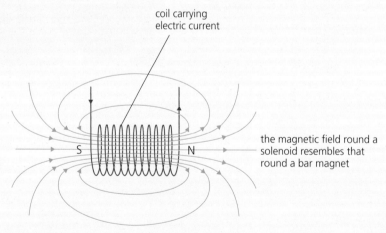

coil carrying electric current

the magnetic field round a solenoid resembles that round a bar magnet

Figure 7.17 Magnetic flux density tells us about the strength of a magnetic field

Force on a wire

The size of the force, *F*, in N on a wire carrying a current at right angles to a magnetic field is proportional to:

- the current, *I*, in A
- the length of wire in the field, *l*, in m.

So *F* is proportional to *Il*. The force is given by the equation:

F = BIl

where *B* is the magnetic flux density, in tesla (T).

Example:

A wire carrying a current of 10 A passes at right angles through a magnetic field of 0.15 T. Calculate the force acting on a wire of length 0.2 m.

F = BIl

B = 0.15 T, *I* = 10 A, *l* = 0.20 m

F = 0.15 × 10 × 0.20 = 0.30 N

2 What happens to the force on a wire carrying a current when:

   a the current in the wire doubles?

   b the current in the wire and the length of wire in the magnetic field both double?

3 Calculate the force on each wire. The wire is perpendicular to the magnetic field.

   a Length of wire is 2 m, current is 0.5 A and the magnetic flux density is 0.05 T.

   b Current in the wire is 2 A, length of the wire is 50 cm and the magnetic flux density is 0.1 T.

4 A wire lies at right angles to a uniform magnetic field of magnetic flux density 0.02 T. 0.3 m of the wire is inside the length of the magnetic field. If the force on this length of wire due to the current in it is 0.03 N downwards, what are the magnitude and direction of the current in the wire?

5 A wire is perpendicular to a magnetic field and has a force on it of 0.05 N. 0.25 m of wire is inside the the magnetic field. Calculate the magnetic flux density when the current in the wire is 3 A.

> **REMEMBER!**
> ....
> Make sure you use the correct units when substituting in the equation *F = BIL*. If the values in the question are not in tesla, amps and metres, you have to change the values to the correct units before starting the calculation.

Forces and magnetic fields

6 Will there be a force acting on a wire carrying a current of 2 A running parallel to a magnetic field? Explain your answer.

7 A metal wire is lying near the equator perpendicular to the Earth's magnetic field. Length of the wire = 2.0 m, weight of wire = 0.30 N, Earth's magnetic flux density B = 0.00003 T.

   a Calculate the current needed to lift the wire off the ground.

   b Explain whether the wire could be lifted if it was placed near one of Earth's magnetic poles.

Electric motors

Learning objectives:

- list equipment that uses motors
- describe how motors work
- describe how to change the speed and direction of rotation of a motor.

KEY WORDS

Fleming's
  left-hand rule
split-ring
  commutator

Every part of a car where you press a button to move something requires a motor.

HIGHER TIER ONLY

Using motors

Electric motors have many uses around the home. Any electrical machine that has moving parts is likely to have a motor. For example, an electric lawn mower, a food processor and an electric drill (Figure 7.18) all contain electric motors.

1 **Give three more examples of equipment that contains an electric motor.**

Figure 7.18 Electric motors are used in many appliances such as this drill

How motors work

When a current passes through a coil placed between the poles of a magnet there is a force on each side of the coil. **Fleming's left-hand rule** shows that the forces on each side of the coil are in opposite directions, so the coil will move. One side of the coil moves up and the other side moves down. This makes the coil start to spin.

A direct current (dc) motor has a **split-ring commutator** that rotates with the coil between two carbon brushes (Figure 7.19). The commutator allows the motor to continue to spin without reversing direction every time it gets to the vertical position.

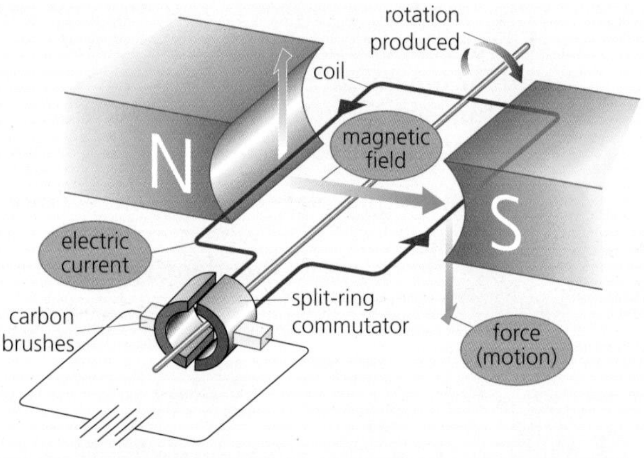

Figure 7.19 The circuit for a d.c. motor

You can use Fleming's left-hand rule to work out the direction of the force. On the left-hand side of the coil in Figure 7.19, the direction of the current is towards the commutator and the field is to the right. So the force is upwards. On the right-hand side of the coil, as the current is in the opposite direction, away from the commutator, the force will also be in the opposite direction, which is downwards. So the coil spins clockwise. When the coil is vertical, the circuit is broken and no current flows. Since the coil is moving, there is enough momentum for the coil to carry on rotating a little further. The side of the coil that was on the left is now on the right, and in contact with the other carbon brush. The direction of the current in the coil then reverses and the force on the other side of the coil is now upwards so the motor continues to spin in the same direction.

Practical motors have curved pole pieces. This produces a radial field (Figure 7.20).

The coil is always at right angles to the magnetic field. This increases the force and keeps it constant as the coil turns.

The direction the motor turns can be reversed by reversing the current or reversing the direction of the magnetic field.

> **KEY INFORMATION**
>
> In an electric motor, the force on a conductor is at right angles to the magnetic field and to the current. For each side of the coil, Fleming's left-hand rule represents the relative orientation of the force, the current in the conductor and the magnetic field.

2 Why does the coil start to spin in Figure 7.19?

3 Explain how a split-ring commutator works.

4 For the rectangular coil in Figure 7.19, only two of the sides experience a force. Why does no force act on the other two sides?

Ac motors

Motors also work with alternating current motors have slip rings instead of a split-ring commutator (Figure 7.21). As the coil rotates, the direction of the current passing through the coil reverses direction. When the coil has gone through half a turn, the direction of the current has reversed.
Look at the left-hand side of the coil. The current flows in the direction AB. The force is downwards and the coil moves anticlockwise. By the time this part of the coil has rotated through half a turn, the direction of the current has reversed and is in the direction CD. The force is upwards so it continues rotating anticlockwise.

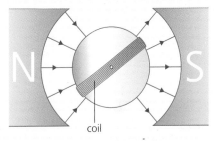

Figure 7.20 A radial field in a motor

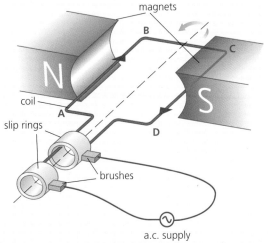

Figure 7.21 An a.c. electric motor

> **KEY INFORMATION**
>
> The factors that affect the size of the force on each side of the coil of wire are:
> • the number of turns on the coil
> • the size of the current
> • the strength of the magnetic field.

5 What effect would there be if both the current and magnetic field were reversed?

6 Give two things that could be done to make the motor turn more slowly.

7 Suggest why you could not use wires to connect the coil of the motor without a commutator.

8 In practical motors, the coil is often wound round some iron. Explain what advantage this gives.

Loudspeakers

Learning objectives:

- describe how a moving-coil loudspeaker works
- compare loudspeakers and headphones.

KEY WORDS

headphones
longitudinal wave
moving-coil
 loudspeaker

Woofers, tweeters and mid-range speakers are types of loudspeakers which sound different because they reproduce different frequencies of sound.

HIGHER TIER ONLY

Producing sounds

A sound wave is a series of oscillations or vibrations that travel in the same direction as the direction of the oscillations – they are **longitudinal waves**. The vibrations travel through the air as pressure waves and transfer energy (Figure 7.22).

MAKING CONNECTIONS

You will need to link the information given here to Chapter 6 on waves, where you looked at longitudinal sound waves in air.

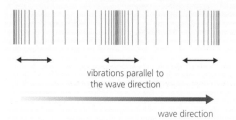

vibrations parallel to
the wave direction

wave direction

Figure 7.22 The oscillations in a sound wave transfer energy

1 **How are sound waves produced?**

The moving-coil loudspeaker

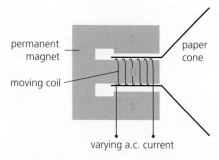

permanent
magnet

paper
cone

moving coil

varying a.c. current

Figure 7.23 The construction of a moving-coil loudspeaker

The loudspeakers in your radio and television are probably like the one shown in Figure 7.23.

A **moving-coil loudspeaker** has a permanent magnet and a coil of wire. The magnet is fixed and the coil can move backwards and forwards. The coil is attached to the cone (usually paper or plastic) of the loudspeaker. The outer edge of the cone is fixed and does not move.

A cylindrical magnet produces a strong radial magnetic field. This is at right angles to the wire in the coil. A signal in the form of a current that varies flows through the coil. A change in the current in the coil causes a change in the magnetic force acting on the coil. The changing force causes the coil to move in and out. As the current varies, the coil moves. As the coil is attached to the cone, the cone also moves, setting up vibrations in the air (Figure 7.24).

2 Explain how an alternating current can make the loudspeaker cone vibrate.

3 Two loudspeakers are connected to the same current. However, one speaker produces louder sound than the other one. Suggest what is different about the louder speaker.

DID YOU KNOW?

A louder sound is produced by the paper cone moving through a larger distance because of a bigger force on the coil.

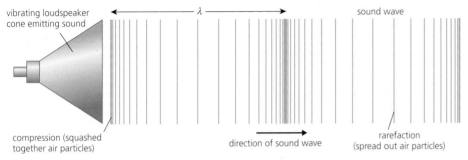

Figure 7.24 Sound waves transfer energy from a loudspeaker

The current flowing in the coil of the loudspeaker causes changes in pressure in the air around the loudspeaker. Energy is transferred from the loudspeaker to the air, making the particles move.

Headphones and loudspeakers

Headphones and loudspeakers work in the same way. The biggest difference between loudspeakers and headphones is size. A loudspeaker needs to carry the sound to a whole room, but the speaker in a headphone only has to move the volume of air inside your ear canal (Figure 7.25).

4 How accurate is it to describe earbud headphones as 'loudspeakers that fit inside your ear'?

5 If you use your mobile phone to play music through earbuds it will do so for several hours but if you decided to use loudspeakers instead, they'd probably need their own batteries. Suggest why.

Figure 7.25 These earbud headphones can fit snugly into the ear

6 Look at Figure 7.23. Explain which way the paper cone moves when the electric current is flowing in at the left-hand side and out on the right-hand side.

The generator effect

Learning objectives:

- describe how a potential difference is induced across the ends of in a wire when it moves in a magnetic field
- identify the factors that affect the size and direction of the induced current or induced potential difference.

KEY WORDS

generator effect
induced
induced current
potential
  difference

When Michael Faraday induced a potential difference in a conductor by changing the magnetic field through a coil he would have had no idea how much his invention would change the world. The effect is used in many things around us.

HIGHER TIER ONLY

Inducing a potential difference

When a wire is held stationary between the poles of a magnet there is no current in the wire. If the wire moves it cuts the magnetic field lines. A potential difference is **induced** across the ends of the wire and if the wire is part of a complete circuit an **induced current** passes along the wire (Figure 7.26). This is called the **generator effect**.

There also is an induced **potential difference** whenever a magnetic field changes near a conductor, even if the conductor does not move (Figure 7.27). Moving a magnet near a conductor changes the magnetic field around the conductor.

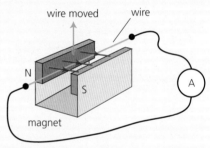

Figure 7.26 Inducing a potential difference by moving a wire in a magnetic field

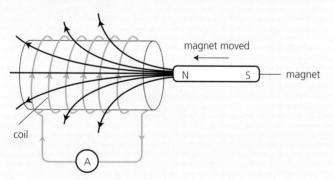

Figure 7.27 Moving the magnet changes the magnetic field through the coil

1. How can a potential difference be induced in a wire in a magnetic field?

2. What is the value of the induced potential difference when the wire is moved parallel to the magnetic field?

Changing the size of the induced potential difference

The size of the induced potential difference depends on the rate at which the magnetic field changes. The induced potential difference can be increased by:

- moving the wire or magnet faster
- using a stronger magnet
- replacing the wire with a coil.

3 A bar magnet is moved quickly into a solenoid. Explain how you can reduce the size of the induced potential difference.

> **KEY INFORMATION**
>
> A potential difference is only induced when something changes. Either the conductor is moving – its position changes – or the magnetic field changes.

What is the direction of the induced potential difference?

The direction of the induced potential difference is always such as to oppose the motion or change producing it.

Energy is transferred from the movement of the magnet into the coil to a flow of current (Figure 7.28).

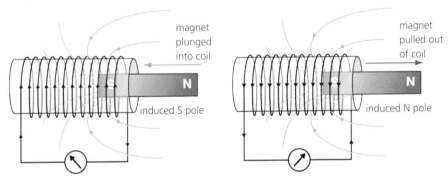

Figure 7.28 Inducing a potential difference in a solenoid

When the bar magnet is moved into the coil, the coil produces a magnetic field because a potential difference is induced and a current flows as there is a complete circuit. The polarity of the magnetic field at the end of the coil closest to the magnet opposes the field of the magnet. This makes it harder to move the magnet into the coil. When you move the magnet out of the coil again, the coil produces a field that opposes the field of the magnet. As the movement is in the opposite direction, the current induced is also in the opposite direction.

The direction of the induced potential difference can be reversed by reversing the direction of the magnetic field; this also reverses the induced current direction. Moving the wire in the opposite direction also reverses the direction of the induced potential difference.

4 Look at Figure 7.28.

   a The blue lines show the magnetic field lines of the magnet. Explain why there is a second magnetic field present.

   b Describe where the poles of this second field occur on both diagrams in Figure 7.28.

   c Explain why it requires a force to move the magnet in and out of the coil.

   d Pretend that the potential difference had been induced in the other direction. Explain what would have happened to the magnet and why energy wouldn't have been conserved.

5 Suggest what you would see if only one end of the meter was connected to the coil.

KEY CONCEPT

The link between electricity and magnetism

Learning objective:

- explore how electricity and magnetism are connected.

KEY WORDS

generator effect
induced potential
  difference
induced current
magnetic field
motor effect

In 1820, Hans Christian Oersted was demonstrating an experiment with batteries and electric current in a circuit. He noticed that every time he switched the batteries on and off, the needle in a compass that was lying on the bench beside his work deflected from magnetic north. This was when he realised that there was a direct relationship between electricity and magnetism.

Michael Faraday discovered a way of producing electricity from magnetism in 1831, which led to the development of the first dc electric motor in 1832.

HIGHER TIER ONLY

Electricity and magnetism

Electricity has to do with the properties and movement of electrically charged particles. Magnetism involves the properties of magnets and magnetic materials.

Electricity and magnetism are interrelated. Movement of electrical charges creates **magnetic fields**, while changes in magnetic fields can generate electricity.

- Current flowing through a circuit creates a magnetic field around the wire.
- Moving part of a wire circuit through a magnetic field **induces a potential difference** across the wire and that potential difference begins to drive electrons round the circuit.

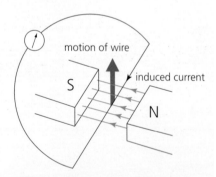

Figure 7.29 One of Faraday's experiments showed that moving a wire through a magnetic field would generate an **induced current** in the wire

1 Name a device where an electric current is used to produce a magnetic field.

2 Name a device where an electric current and a magnetic field produce motion.

3 Name a device where motion in a magnetic field produces an electric current.

Explaining the motor effect

Imagine you put a current-carrying conductor, which has a magnetic field around it, into a magnetic field. The magnetic field around the conductor interacts with the magnetic field of the magnet. The combination of the magnetic fields creates an instantaneous force between the magnet and the conductor. This is the underlying principle behind the electric motor.

Figure 7.30 A modern electric motor in a disc drive

4 Describe in your own words the underlying principle of the electric motor.

5 A wire carrying a current is placed between the poles of a U-shaped magnet (Figure 7.31). Describe how the direction of the resulting magnetic force on the wire can be found.

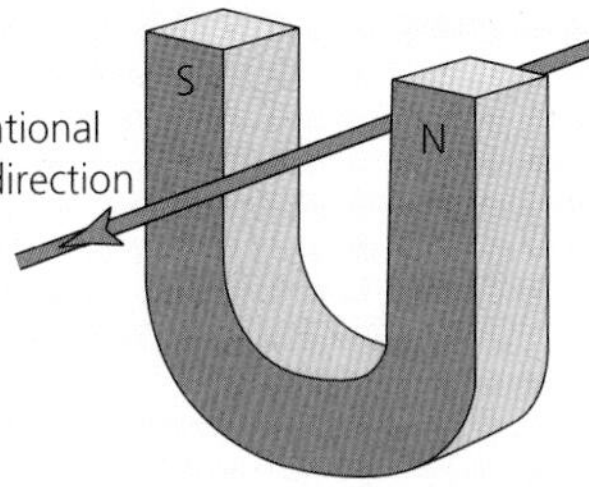

conventional current direction

Figure 7.31

Electromagnetic induction

The effect that Michael Faraday discovered is called electromagnetic induction. A wire moving through a magnetic field induces a potential difference which pushes current around a circuit. This **generator effect** is the underlying principle of the electric generator.

How does moving a wire through a magnetic field create an induced potential difference? Imagine you could see the invisible lines of magnetic force running north to south between the poles of a horseshoe magnet. A wire moving at right angles to the magnetic field cuts through magnetic field lines. By changing the magnetic field around the wire, a potential difference is induced across the ends of the wire. The effect would be exactly the same if you kept the wire still and moved the magnet. The key point is that there has to be relative motion between the conductor and the magnetic field.

6 Describe the differences and similarities between motors and generators.

> **REMEMBER!**
>
> The dc generator or dynamo is the same as the dc motor in design. However, in a motor the electric current transfers energy to kinetic energy, and in the dynamo energy is transferred from the movement of the magnet as a flow of current.

Using the generator effect

Learning objectives:

- explain how moving-coil microphones use the generator effect
- explain how a dynamo generates direct current and an alternator generates alternating current
- for a dynamo and alternator, draw and interpret graphs of potential difference generated in the coil against time.

KEY WORDS

alternator
dynamo
moving-coil
  microphone

In a bicycle dynamo a permanent magnet is rotated within a coil rather than the coil rotating between the poles of a permanent magnet.

HIGHER TIER ONLY

Moving-coil microphone

Moving-coil microphones act like loudspeakers in reverse. Energy is transferred from the moving air into the coil, causing a current to flow.

The construction of a moving-coil microphone (Figure 7.32) is very similar to a loudspeaker. There is a coil connected to a diaphragm. The edges of the diaphragm are fixed. The coil surrounds a permanent magnet. Sound waves make the diaphragm vibrate. These vibrations make the coil move backwards and forwards in the gap, inducing a potential difference and causing a varying current to flow.

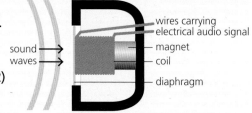

Figure 7.32 A moving-coil microphone

1. Explain how a microphone is similar to a loudspeaker.
2. Explain how a moving-coil microphone converts the pressure variations in sound waves into variations in current.
3. Suggest why the coil has a very large number of turns of very thin wire.

The dynamo

Figure 7.33 shows a **dynamo**. It is used to generate a direct current by moving the coil in a magnetic field. As the coil turns it cuts the magnetic field lines. Doing this induces a potential difference between the ends of the coil. This causes an induced current to pass through the coil and through the circuit the coil is a part of.

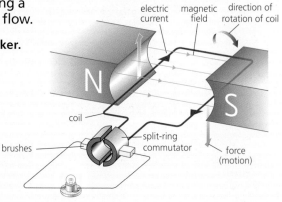

Figure 7.33 A dc dynamo

The split-ring commutator rotates with the coil and maintains a complete circuit with the external circuit. However, each half turn the sides of the coil connect to the opposite side of the circuit. So although the direction of the induced potential difference between the ends of the coil reverses when the coil cuts through field lines in the opposite direction, the direction of the current induced in the circuit does not change.

The ac alternator

An **alternator** also uses the generator effect. As with the dynamo, the size of the induced current can be increased by:

- increasing the number of turns on the coil
- using more powerful magnets
- turning the coil faster.

In an alternator, slip rings are connected to the ends of the coil to allow the coil to spin without winding the wire around itself. The brushes are conducting contacts that touch the slip rings and complete the circuit (Figure 7.34).

⑤ **Explain the difference between the construction of a dc dynamo and an ac generator.**

As the coil rotates, cutting magnetic field lines, it generates a potential difference between the ends of the coil. The potential difference is greatest at the instant when the coil is horizontal, because this is when the moving coil is cutting through field lines (Figure 7.35 A).

When the coil is vertical and at that instant moving parallel to the magnetic field direction, no field lines are being cut (Figure 7.35 B) so the potential difference is zero.

As the coil continues to rotate, the side that was coming up through the field is now going down (Figure 7.35 C), so the induced potential difference is negative.

This means that as each side of the coil moves down and then up through the magnetic field it makes current pass first in one direction and then the other. This is alternating current (ac) (Figure 7.36).

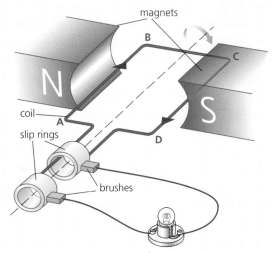

Figure 7.34 An ac alternator

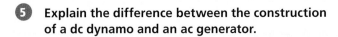

Figure 7.35 The rotating coil of an alternator at different positions. The magnetic field direction is shown in green

Figure 7.36 The output of an alternator

⑥ **Explain how an alternating potential difference is induced.**

⑦ **Suggest what increasing the speed of rotation will do to the potential difference generated.**

⑧ **Explain why the potential difference generated is zero at various points of the coil's rotation.**

⑨ **Give two ways of increasing the induced current generated in an alternator.**

⑩ **The light bulb in Figure 7.34 blows. Suggest why it is now easier to turn the coil.**

Transformers

Learning objectives:

- explain how a transformer both uses and produces alternating current
- explain the relationship between the number of turns in the primary coil and the number in the secondary coil
- calculate the current that needs to be provided to produce a particular power output.

KEY WORDS

efficiency
step-down transformer
step-up transformer
transformer

The first power distribution system was in New York in 1882. It ran on 110 V dc Since it could not use transformers, it could only supply a very small area due to the inefficiency of power transfer.

HIGHER TIER ONLY

Designing a transformer

A **transformer** changes the size of an alternating potential difference. It consists of two coils of wire, wound separately on an iron core (Figure 7.38). An alternating potential difference applied to the primary coil produces an alternating current in the primary coil. This produces an alternating magnetic field in the iron core. The strength of the magnetic field is increased because the iron is easily magnetised. The alternating magnetic field in the iron core induces a potential difference in the secondary coil. This induces an alternating current in the circuit connected to the secondary coil.

A transformer that increases the potential difference is called a **step-up transformer**; it has more turns on the secondary coil than on the primary coil. One that decreases the potential difference is called a **step-down transformer**; it has more turns on the primary coil than on the secondary coil.

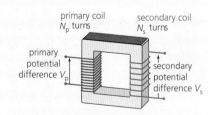

Figure 7.37 A mobile phone charger contains a transformer

1. Would you use a step-up or a step-down transformer to:
   a change 25 kV to 400 kV?
   b change 240 V from a mains socket to 19 V to charge a laptop?
2. Why can't you use a transformer to change the potential difference of a battery?
3. Explain how induction is used in a transformer.

Turns ratio

The ratio of the potential differences across the primary and secondary coils of a transformer, V_p and V_s (Figure 7.38), depends on the ratio of the number of turns on each coil, n_p and n_s:

$$\frac{V_p}{V_s} = \frac{n_p}{n_s}$$ The potential differences are measured in volts (V).

Example: A transformer changes 240 V from the mains electricity supply to 5 V to charge a mobile phone. There are 720 turns on the primary coil; how many turns must there be on the secondary coil?

Figure 7.38 The ratio of the potential differences across the primary and secondary coils is equal to the ratio of the number of turns

Using $V_p/V_s = n_p/n_s$ and substituting values, 240 V ÷ 5 V = 720 turns ÷ n_s
Therefore n_s = 720 × 5/240 = 15 turns.

4. To change 240 V a.c. to 12 V a.c., how many turns should there be on the secondary if there are 720 turns on the primary coil?

5. To change 240 V a.c. to 20 V a.c., how many turns should there be on the primary coil if there are 100 turns on the secondary coil?

6. If you fed 240 V a.c. into a 100 turn primary coil and had 10 turns on the secondary coil, what would the potential difference across the secondary coil be?

Power input and output in a transformer

Efficiency compares the power output from a machine with the input. If a transformer is 100% efficient, the power output is equal to the power input. Electrical power is calculated from current × potential difference, so we can write this as:

$$V_s \times I_s = V_p \times I_p$$

where $V_s \times I_s$ is the power output (secondary coil), $V_p \times I_p$ is the power input (primary coil) and power is measured in watts (W). We can use this to calculate the current drawn from the input supply to provide a particular power output.

Example:
If the output of a mobile phone charger is 5 V and 2 A and it works from 230 V ac mains supply., how much current does it draw from the electricity supply?

Power output, $V_s \times I_s$ = 5 V × 2 A
= 10 W

Assuming the transformer is 100% efficient, the power input will be the same as the power output, 10 W.

Power input $= V_p \times I_p = 230 \text{ V} \times I_p$

Therefore, I_p $= \dfrac{10 \text{ W}}{230 \text{ V}}$

$= 0.043 \text{ A}$

A step-down transformer produces a lower potential difference, so the current is greater. Similarly, a step-up transformer produces a higher potential difference so the current leaving is less than it was in the primary.

For example, if a 3 kV generator fed 20 A into a step-up trans-former to run a power distribution system at 60 kV, the current output would only be 1 A. At lower current levels the energy losses from the power lines are much less and so the power would be distributed much more efficiently.

7. If the output of a laptop charger is 19 V and 3.4 A and the charger runs from 240 V a.c. how much current does it draw?

8. In the example above:

    a Show that the current output would be 1 A.

    b If the step up transformer has 500 turns on the primary coil, how many turns must be on the secondary?

REMEMBER!

Remember that step-up transformers have more turns on the secondary, which produces a greater potential difference and a smaller current. Step-down transformers do the opposite.

MAKING LINKS

The National Grid is a system of cables and transformers linking power stations to consumers (see Figure 2.39 in chapter 2). Remember, the higher the current, the more energy is wasted and the National Grid becomes less efficient.

MATHS SKILLS

Rearranging equations

Learning objective:

- change the subject of a equation.

Equations are used throughout Physics. Calculating the value of a from the equation $a = b \times c$ when you know b and c is straightforward. It becomes more difficult when you want to work out the values of b or c. You need to rearrange the equation.

HIGHER TIER ONLY

Force on a conductor

The force on a conductor carrying a current is $F = BIL$, where the force (F) is in newtons, N, B is the magnetic flux density, in tesla, T, the current (I) flowing through the conductor is in amps, A and the length (L) of the conductor is in metres, m.

To work out the force when you know B, I and L, **substitute** the values into the equation. To work out one of B, I or L you need to **rearrange** the equation. You can rearrange the equation and substitute the values or substitute the values and rearrange the equation.

Example:

A conductor 50 m long, carrying a current of 5 A, is placed in a magnetic field. Calculate the magnetic flux density when a force of 0.02 N is produced on the conductor.

Method 1: substitute the values and rearrange the equation to make B the **subject of the equation**:

$F = BIl$

$0.02\,\text{N} = B \times 5\,\text{A} \times 50\,\text{m}$

Rearrange the equation: divide both sides by 5 A $\times$ 50 m:

$B = \dfrac{0.02\,\text{N}}{(5\,\text{A} \times 50\,\text{m})}$

$\quad = 8 \times 10^{-5}\,\text{T}$

Remember the unit in your answer.

MATHS

Always do the same operation to both sides of the equation.

Method 2: rearrange the equation and substitute the values:

$F = BIl$

Rearrange the equation: divide both sides by Il:

$B = \dfrac{F}{Il}$

Substitute the values into the rearranged equation:

$$B = \frac{0.02\,\text{N}}{(5\,\text{A} \times 50\,\text{m})}$$

$$= 8 \times 10^{-5}\,\text{T}$$

1. Calculate the force on a conductor that is 25 m long carrying 3 A in a magnetic field of flux density 3×10^{-5} T.

2. a Calculate the current in a coil of wire 250 m long with a magnetic field of flux density 1.0×10^{-4} T, when the force is 0.125 N. Work out the answer using both methods in the example.

   b Which method did you find easier?

3. Calculate the length of conductor when a conductor carrying a current of 2.5 A experiences a force of 0.05 N in a magnetic field of flux density 3×10^{-5} T. Use your preferred method.

Transformer equation

The relationship between primary and secondary coils is expressed as:

$$\frac{V_p}{V_s} = \frac{n_p}{n_s}$$

where V_p is the potential difference across the primary coil; V_s is the potential difference across the secondary coil, n_p is the number of turns in the primary coil; n_s is the number of turns in the secondary coil.

With this equation you will always have to rearrange it.

Example:

Work out the potential difference across the secondary coil when the potential difference across the primary coil is 230 V, the number of turns on the primary coil is 400, and the number of turns on the secondary coil is 20.

$$\frac{V_p}{V_s} = \frac{n_p}{n_s}$$

Multiply both sides by V_s: $\quad V_p = \dfrac{n_p \times V_s}{n_s}$

Multiply both sides by n_s: $\quad V_p \times n_s = n_p \times V_s$

Divide both sides by n_p: $\quad \dfrac{V_p \times n_s}{n_p} = V_s$

Work out the value of V_s: $\quad V_p = 230\,\text{V}; n_s = 20; n_p = 400$

$$\frac{230\,\text{V} \times 20}{400} = 11.5\,\text{V}$$

Check your progress

You should be able to:

Recall that like poles repel, unlike poles attract → Recognise that the poles of a magnet are the places where the magnetic forces are strongest → Explain that an induced magnet is a material that is only magnetic when it is placed in a magnetic field

Plot the magnetic field around a bar magnet → Recognise that the magnetic field is the region around a magnet where a force acts on another magnet or on a magnetic material → Explain how the behaviour of a magnetic compass is related to evidence that the core of the Earth must be magnetic

State how the strength of an electromagnet can be increased → Draw the magnetic field around a conducting wire and a solenoid → Explain how electromagnets are used in devices

State that a force acts on a current-carrying conductor in a magnetic field → Describe the motor effect that applies to a current-carrying conductor in a magnetic field → Explain how the direction of the force on the conductor can be identified using Fleming's left-hand rule

State that magnetic flux density is measured in tesla (T) → Explain what the size of a force on a conductor depends on → Use the equation $F = BIL$ to calculate the force on a conductor

State what the generator effect is → Recognise that the size of the induced potential difference (and so induced current) can be increased by increasing the speed of movement or by increasing the strength of the magnetic field → Explain that if the direction of motion of the conductor or the polarity of the magnetic field is reversed, the direction of the induced potential difference and any induced current is reversed

State that a dynamo generates direct current and an alternator generates alternating current → Describe that when a coil is rotated in a magnetic field an alternating current is induced in the coil

Draw graphs of potential difference generated in the coil against time → Explain how the generator effect is used in a dynamo to generate dc with the use of commutator; and is used in an alternator to generate ac

Describe how to draw and interpret graphs of potential difference generated in the coil against time

State what a basic transformer consists of and the difference between step-up transformer and a step-down transformer → Explain how the potential differences across the two coils depend on the number of turns on each coil and how the potential difference is induced → Use and apply the expression $\dfrac{V_p}{V_s} = \dfrac{n_p}{n_s}$

State that high potential differences are used to reduce power transmission losses → Describe how when the potential difference is increased the current decreases for the same power transmitted → Explain how power transmission losses are related to the square of the current

Worked example

① How can you increase the strength of the magnetic field created by a current through a wire?

To increase the strength of the magnetic field you need to coil the wire.

Correct, but that is only part of the answer. There are two more ways in which you can strengthen the magnetic field: increase the current and wrap the coils around an iron core.

② Explain how continuous rotation is produced in an electric motor.

When a wire carrying a current is put into a magnetic field, one force reacts to the other and this pushes the wire down. If this is a coil then it begins to go round.

You should aim to show that you really understand exactly what is going on and where the forces are coming from.

Describe the interaction between the magnetic field in the wire and the magnetic field of the magnet.

With a complicated answer like this, it can be helpful to draw a diagram. In this case the diagram would show the coil, magnetic field and direction of movement.

③ The primary coil of a transformer has 75 turns and a potential difference of 2.5 V. Calculate the potential difference in the secondary coil when there are 225 turns in the secondary coil.

$$\frac{V_p}{V_s} = \frac{n_p}{n_s}$$

$$\text{So } V_s = \frac{V_p \times N_p}{N_s}$$

$$= \frac{2.5 \times 75}{225}$$

$$= 0.83 \text{ volts}$$

The student started with the right equation and showed their working, but the answer is incorrect. They have rearranged the equation for V_s incorrectly: n_p and n_s are the wrong way round.
Check that you can rearrange equations correctly.

End of chapter questions

Getting started

1 Where is the magnetic field around a bar magnet strongest?

  a near the poles   b along the side

  c above the magnet   d well away from the magnet   `1 Mark`

2 Which of these will not strengthen the magnetic field created by a current through a wire?   `1 Mark`

  a shape the wire into a coil   b stripping the plastic insulation off the wire

  c increasing the potential difference   d using a thicker wire

3 Give two differences between a permanent magnet and an induced magnet.   `2 Marks`

4 What happens when two like poles are brought near to each other?   `1 Mark`

  a they attract   b there is no interaction

  c they repel   d it isn't possible as they are at opposite ends of the same magnet.

5 There are two small iron bars on the bench. One is magnetised and the other is not. Explain how you could identify which is which.   `2 Marks`

6 Jemima has made an electromagnet using a coil of wire and a power supply. Which of these will not indicate how strong the magnetic field is?

  a seeing from how far away the coil will attract a paper clip

  b seeing how many paper clips it will attract

  c seeing what potential difference the power supply is set to

  d seeing from how far away it can affect a plotting compass   `1 Mark`

7 An electric doorbell has an electromagnet in it. This is to:

  a make the doorbell heavier   b make the doorbell louder

  c attract the striker to hit the gong   d use more electricity   `1 Mark`

8 A plotting compass is placed near a bar magnet. Which direction does the arrow of the plotting compass point?

  a towards the Earth's magnetic north pole   b towards the Earth's magnetic south pole

  c towards the magnet's north pole   d towards the magnet's south pole   `1 Mark`

Going further

9 A moving coil microphone works because:

a electricity makes the coil move

b electricity makes a noise

c the moving coil makes a noise

d pressure differences due to sound make the coil move `1 Mark`

10 Which of these will not change the direction of the force in the motor effect?

a increase the potential difference

b reverse the potential difference

c reverse the polarity of the magnets

d swap the contacts on the power supply `1 Mark`

11 Describe how you can make an electromagnet as strong as possible. `2 Marks`

12 A model railway requires 12 V ac to power the engine. Calculate:

a the number of turns on the secondary coil of the transformer if the primary coil has 500 turns and runs from 240 V ac

b the current flowing in the primary coil if the railway needs 1.5 A. `4 Marks`

13 A student was investigating how the magnetic force on a length of wire depends on its length. Explain how the student can make this a fair test. `2 Marks`

More challenging

14 Which one of these does the size of the force in the motor effect not depend upon?

　　a The strength of the magnetic field　b How hard the person pushes the wire

　　c The size of the current　　　　　　d The length of the wire

1 Mark

15 A portable radio uses a lower potential difference than the mains. When designing a transformer to provide it with power, which of these is true?

　　a There should be fewer coils on the primary coil than on the secondary.

　　b There should be the same number of coils on the primary coil as on the secondary.

　　c There should be fewer coils on the secondary coil than on the primary.

　　d The number of coils doesn't make any difference.

1 Mark

16 The diagram shows a moving-coil loudspeaker. Explain how it works.

2 Marks

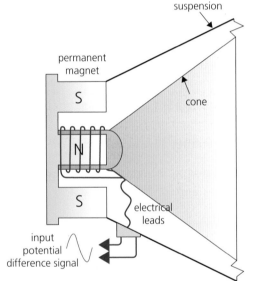

17 A wire is wound into a solenoid and the solenoid is connected to a datalogger. The datalogger records the potential difference across the solenoid and plots this as a graph of potential difference against time.

The diagram shows the graph obtained by the datalogger when a bar magnet was made to move through the solenoid.

Use ideas about the generator effect to explain the shape of the graph. You should also refer to the speed of the magnet.

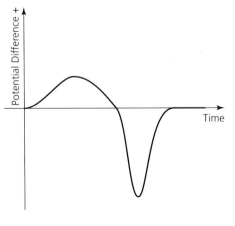

6 Marks

Most demanding

18 Explain how the direction of the induced potential difference in the generator effect can be reversed.

2 Marks

19 The diagram shows the potential difference generated by a coil moving through a magnetic field. Explain how the shape of the waveform matches the position of the coil.

2 Marks

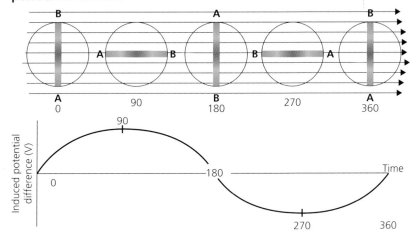

20 A generator on an island produces electricity at 750 V and 12 A. The power distribution lines have a resistance of 50 Ω. If the power loss in the lines is given by I^2R, determine how much power would be saved by transforming it to 2.25 kV before transmission. Assume the transformers are 100% efficient.

6 Marks

Total: 40 Marks

SPACE

OUR SUN IS A STAR

- The Sun is the star at the centre of our Solar System.
- The Earth is one of eight planets orbiting the Sun.
- The Sun is a star in a galaxy called the Milky Way.
- There are many other stars and galaxies in the Universe.

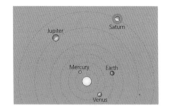

THE POSITION AND MOVEMENT OF THE EARTH AFFECTS OUR DAYS AND SEASONS

- The Earth turns on its axis once every 24 hours.
- The half of the Earth facing towards the Sun has daylight and the half facing away from the Sun has night.
- The Earth's axis is tilted.
- The part of the Earth tilted towards the Sun has more hours of daylight and the part tilted away from the Sun has fewer hours of daylight.

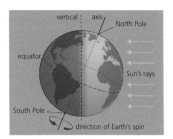

WE USE ENORMOUS UNITS TO MEASURE DISTANCE IN SPACE PHYSICS

- A light year is the distance light travels in one year.
- In astronomy we measure distances in light years.
- The distances involved in astronomy are enormous.

THE FORCE OF GRAVITY IS IMPORTANT TO THE EARTH

- Weight = mass × gravitational field strength (g).
- On the Earth, gravitational field strength (g) = 10 N/kg.
- Gravitational field strength (g) is different on other planets and stars.
- The force of gravity keeps the Moon in orbit around the Earth and the Earth in orbit around the Sun.

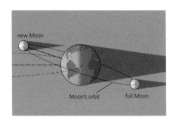

WHAT CAN WE LEARN ABOUT STARS?

- Our Solar System contains planets and dwarf planets which orbit the Sun and natural satellites called moons which orbit the planets.
- Fusion processes in stars lead to the formation of new elements.
- A star goes through a life cycle which is determined by the size of the star.

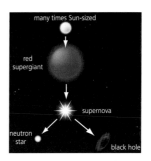

WHAT MOVEMENTS CAN WE DETECT IN SPACE PHYSICS?

- Moons orbit many of the planets in the Solar System.
- Artificial satellites have also been placed in orbit around the Earth.
- The Universe is expanding at an increasing rate.

WHAT DO OUR MEASUREMENTS TELL US ABOUT THE UNIVERSE?

- The increase in the wavelength called red-shift tells us that most stars and galaxies are moving away from us.
- The further away the galaxy, the faster it is moving and the bigger the red-shift.
- Red-shift provides evidence that the Universe is still expanding and supports the Big Bang theory.
- There is still much about the Universe that is not understood, such as dark matter and dark energy.

WHAT IS THE ROLE OF GRAVITY IN SPACE PHYSICS?

- Our Sun and other stars are formed from dust and gas (nebulae) pulled together by the force of gravity.
- Gravity allows planets and satellites to maintain their orbits.
- The force of gravity acts towards the centre of the orbit and causes acceleration in that direction which results in a changing velocity but unchanged speed.

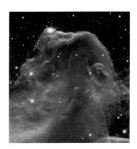

The Solar System

Learning objectives:

- describe the orbits of planets and moons in the Solar System
- distinguish between planets, dwarf planets and moons.

KEY WORDS

dwarf planet
natural satellite
orbit
planet
Solar System

Scientists have discovered that there is water on Mars. They estimate that both the North and South polar ice caps hold five million cubic kilometres of frozen water. The deep river channels, deltas and lake beds indicate that water has flowed on Mars in the past.

Figure 8.1 The side of the Newton Crater on Mars. The features could be evidence of water on Mars

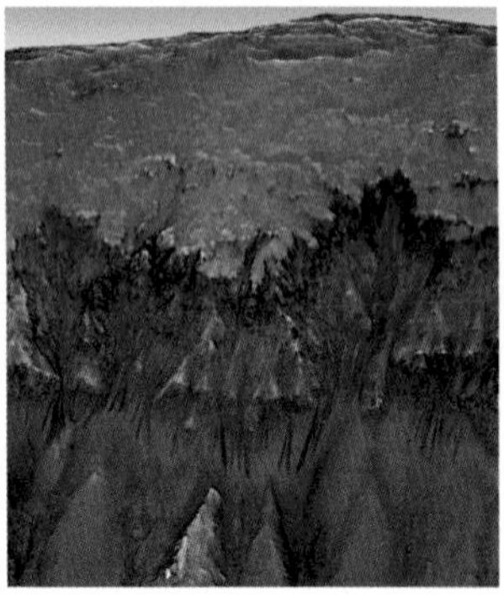

The planets in the Solar System

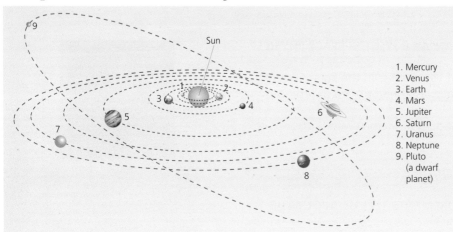

Sun
1
2
3
4
5
6
7
8
9
1. Mercury
2. Venus
3. Earth
4. Mars
5. Jupiter
6. Saturn
7. Uranus
8. Neptune
9. Pluto (a dwarf planet)

Figure 8.2 The Solar System

The Sun is at the centre of our **Solar System**. Eight **planets orbit** the Sun. The Earth is the third planet from the Sun, between Venus and Mars.

1. **Which planet is closest to the Sun?**
2. **Describe what the Sun would look like if you stood on the surface of Pluto.**

The four inner planets, Mercury, Venus, Earth and Mars, are made of rock and are relatively small (Figure 8.3).

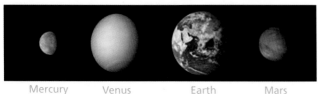

Mercury
Venus
Earth
Mars

Figure 8.3 The inner planets: Mercury, Venus, Earth and Mars

Mercury is the smallest planet and is closest to the Sun. Venus is about the same size as the Earth. Venus has an atmosphere, and this traps energy radiated from the Sun, making Venus the planet with the highest average temperature.

3 **Figure 8.2 shows the orbits drawn to scale. Why have the planets and the Sun been drawn to a much bigger scale?**

4 **Name the largest planet in the Solar System.**

The four outer planets, Jupiter, Saturn, Uranus and Neptune, are much larger, are made up of gas and are called gas giants.

Figure 8.4 The four outer planets

Dwarf planets and moons

Dwarf planets in the Solar System are small planets, mostly in the outer Solar System, beyond the orbit of Neptune. Pluto is classed as a **dwarf planet**.

Moons are **natural satellites** of a planet or dwarf planet.

5 **Is Pluto a planet, dwarf planet or moon?**

The differences between planets, dwarf planets and moons

Planets are clearly different from stars, because stars, like our Sun, make their own light and planets do not. Our planet, the Earth, is in orbit around a star. What is the difference between planets and dwarf planets, and moons (like our Moon)?

To be a planet an object has to be spherical, in orbit around a star and have a gravitational field strong enough to clear other material out of its orbit. A dwarf planet is spherical and orbits a star but can't clear other material out of its own orbit because its gravitational field is too weak. A moon orbits a planet and not a star.

6 **Why might one planet have a stronger gravitational field than another?**

7 **What happens to bits of rock and dust in space that end up in the Earth's gravitational field?**

8 **Give three reasons why Pluto is very difficult to see from the Earth.**

Figure 8.5 This image of Pluto was taken by the New Horizons spacecraft in July 2015

DID YOU KNOW?

Although Pluto is tiny, it has at least five moons. The largest one, Charon, is half the diameter of Pluto, and this affects the movement of the other four moons.

Orbits of planets, moons and artificial satellites

Learning objectives:

- compare the orbital motion of moons, artificial satellites and planets in the Solar System
- describe what keeps bodies in orbit around planets and stars

- explain how, for circular orbits, an object can have a changing velocity but unchanged speed
- explain why bodies must move at a particular speed to stay in orbit at a particular distance.

KEY WORDS

artificial satellite
natural satellite
orbit
vector

The Sun is huge. It contains about 98% of the mass and material of our Solar System. The huge mass of the Sun provides the force of gravity that keeps our Solar System together and in orbit around it.

Orbits of the planets

The planets in the Solar System move around the Sun in almost circular **orbits**. The table gives information about the planets in the Solar System.

Planet	Average distance from the Sun in million km	Time to orbit the Sun
Mercury	57	88 days
Venus	108	225 days
Earth	150	1 year
Mars	228	1.9 years
Jupiter	778	11.9 years
Saturn	1429	29.5 years
Uranus	2870	84 years
Neptune	4500	165 years

Figure 8.6 An astronaut's footprint on the Moon

1 What force keeps the planets in orbit around the Sun?

2 Describe the relationship between the distance of a planet from the Sun and the time it takes to orbit the Sun.

3 Suggest a reason why the relationship in question 2 is like this.

The Earth's Moon

The Moon is the Earth's **natural satellite** and orbits the Earth every 27.3 days. We always see the same face of the Moon because it completes one orbit of the Earth in the same time as it takes to complete one rotation.

DID YOU KNOW?

When astronauts look at the Earth from space they can see it change shape in a similar way to the way the Moon changes shape each month.

Figure 8.7 View of the Earth from the Moon taken on the Apollo 8 Mission

4 What force keeps the Moon in orbit around the Earth?

5 Draw a diagram to show:

a the path a planet would take around the Sun

b the path a moon of the planet would take.

Artificial satellites

There are many **artificial satellites** in orbit around the Earth. They have been put there for a variety of reasons, such as looking at weather patterns, broadcasting TV signals, spying and space exploration. One satellite we can occasionally see quite easily from the Earth is the International Space Station.

Artificial satellites are used for communications, GPS, weather forecasting, surveys of the Earth's surface and map making.

Figure 8.8 Sputnik 1 was the first artificial satellite launched, in 1957

HIGHER TIER ONLY

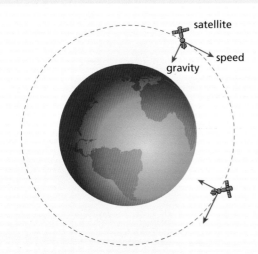

Figure 8.9 How a gravitational field keeps objects in orbit

A satellite in a stable orbit moves at a constant speed in a circular orbit at a particular distance from the object it is orbiting. Remember that **vector quantities**, such as velocity, have both magnitude and direction. An artificial satellite may be orbiting the Earth at a constant speed but its direction of motion keeps changing so its velocity is constantly changing.

When an artificial satellite is put into orbit it is taken up to a certain height above the Earth's surface and launched at right angles to the direction that gravity is pulling it back towards the Earth (Figure 8.9). Instead of flying off into space, gravity pulls it into an orbit. The launch speed is crucial. Too great and the satellite will escape the Earth's gravitational attraction; too small and it will fall to Earth.

Sometimes a satellite needs to be in a low orbit. Some weather satellites, for example, need to be closer to the Earth to get the pictures needed for forecasting. If the radius of the orbit is less, the speed has to be greater to keep the satellite in a stable orbit.

6 How can a satellite be changing velocity but staying at the same speed?

7 What factors affect the stability of the orbital motion of an artificial satellite?

8 Describe how the speed of an artificial satellite is linked to the height of the satellite above the Earth.

The Sun and other stars

Learning objectives:

- describe how the Sun and other stars formed
- describe the nuclear fusion reactions in the Sun.

KEY WORDS

galaxy
nebula
nuclear fusion
protostar

The Earth is approximately 150 million km from the Sun and it takes light from the Sun eight minutes to reach the Earth.

Stars and galaxies

Our Sun is our nearest star. It is just one of billions of stars that make up our **galaxy**, the **Milky Way**. The Universe is made up of billions of galaxies.

DID YOU KNOW?

The Milky Way rotates around the central core of the galaxy. Astronomers believe that there is a black hole at the centre of the galaxy.

1 Put these bodies in order of size. Put the smallest object first: **star, planet, galaxy, Solar System**

The origin of stars

Space is not just empty vacuum. Space contains huge clouds of dust, hydrogen and helium gas which are called **nebulae** (Figure 8.11). These are sometimes called the birthplaces of stars.

Figure 8.10 Spiral galaxy M74, taken from the Hubble Space Telescope

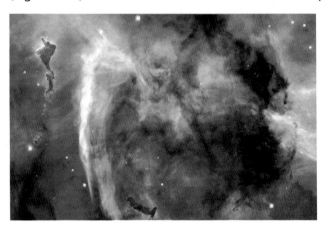

Figure 8.11 The Carina nebula

Over millions of years, the force of gravity pulls the gas and dust in the cloud together. As the vast cloud compresses it get hotter and hotter until the core begins to glow and it becomes a **protostar**.

The protostar's gravitational field continues to attract more gas and dust, and the protostar becomes larger and hotter. Eventually the core becomes so hot that the nuclei of atoms of hydrogen gas begin to fuse, which triggers the release of huge amounts of energy. At this point the core of the protostar ignites in a massive nuclear reaction. This is the moment when a star is formed (Figure 8.12).

Figure 8.12 Artist's rendering of the birth of a new star

2. State the materials a star is formed from.

3. States the force that pulls these materials together.

4. Explain why it takes millions of years to form a protostar.

5. Describe how a star is formed.

Nuclear fusion

In Chapter 4 you learned that **nuclear fusion** occurs when two hydrogen nuclei join or fuse together to form a new nucleus of helium (Figure 8.13). This is only one of the fusion reactions in the Sun that release energy. There are many other reactions as well.

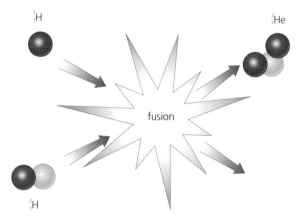

Figure 8.13 Fusion of hydrogen nuclei in the Sun is one of the fusion reactions that release energy

Since the nuclei are positively charged they repel each other. The energy needed to overcome the repulsion comes from the high temperatures created by the force of gravity compressing the gases.

At 15 000 000 °C the hydrogen nuclei move with such high speeds that they can fuse.

6. What element is the Sun mostly made up of?

7. Why do fusion reactions require extremely high temperatures?

8. Explain what happens in the nuclear fusion reactions inside the core of the Sun.

9. Suggest whether there is any helium in a protostar.

10. Why is it wrong to say that hydrogen is being burned in our Sun?

During periods of high solar activity, the Sun can eject massive amounts of charged particles (plasma) in solar flares. Sometimes, solar flares (Figure 8.15) can be so powerful that they can cause satellites in orbit around Earth to malfunction.

DID YOU KNOW?

We normally measure temperature in degrees Celsius. In physics we use the kelvin scale. 0 K = −273 °C

DID YOU KNOW?

The Sun releases about one billion kilograms of charged particles into space each second. This creates a solar wind that creates the Northern Lights (Aurora Borealis) when these particles reach the Earth's atmosphere (Figure 8.14).

Figure 8.14 The Aurora Borealis over Greenland

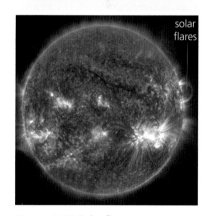

Figure 8.15 Solar flare.

Main sequence of a star

Learning objectives:

- describe the main sequence stage of a star's life cycle
- identify the forces that are in equilibrium in a stable star.

Even though the Sun is in a stable state, its surface is very turbulent. We can see this turbulence (called sunspots) by observing the different wavelengths of the electromagnetic radiation the Sun emits.

Main sequence stars

Stars spend most of their life cycle in a stable state. In this state the stars are called **main sequence** stars. The Sun is a main sequence star. The surface temperature of a main sequence star and how much radiation it emits varies depending on its size.

1. What is a main sequence star?

2. Suggest why most stars in the night sky are main sequence stars.

Forces in equilibrium

In main sequence stars, the forces inside the star are in **equilibrium** (Figure 8.17). The force of gravity acting on all the matter inside the star is trying to collapse the star into a smaller space. Fusion is taking place in the core and the energy released has to move out of the core. The enormous temperatures caused by the fusion reactions produce an outwards force. The force pushing outwards due to the fusion reactions and the force of gravity pulling back into the core are balanced and so the star is stable.

Each star spends most of its life in this stable state.

Figure 8.16 High frequency ultraviolet and X-ray photographs of the Sun

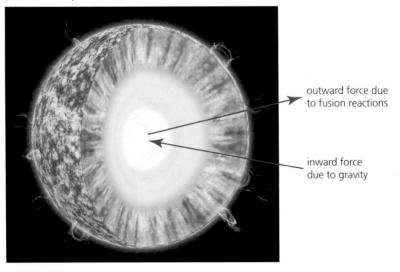

outward force due to fusion reactions

inward force due to gravity

Figure 8.17 In a stable star the gravitational force acting inwards balances the outwards force caused by fusion reactions

3 What does 'forces in equilibrium' mean?

4 What are the characteristics of a main sequence star?

5 Which gas is our Sun making lots of at the moment?

6 Explain why the size of a main sequence star is stable.

The life cycle of stars

The Sun is a main sequence star of average size. It is about halfway through its 10 billion year life – it has used up about half of its store of hydrogen by fusing it to helium. The temperature at the surface of the Sun is about several thousand degrees Celsius, but its core temperature is 15 million degrees Celsius. An average star like our Sun lasts for about 10 billion years, until all of the hydrogen in the core has fused to form helium. Then the balance between the gravitational force inwards and the outwards force due to the fusion reactions changes, and the helium core starts to contract. The temperature increases until it is hot enough to fuse helium to create larger (heavier) elements.

What happens now depends on the size of the star.

DID YOU KNOW?
....

In astronomy the mass of the Sun is used as a base unit to compare other stars against. The mass of our Sun is referred to as 1 solar mass.

7 At what point in a star's life cycle is a star:

   a contracting?

   b at a stable size?

8 Here is some data for some main sequence stars:

Star	85 Ceti	Eta Aurigae	Gliese 758	Kepler 42
Mass (solar masses)	2.4	5.0	0.97	0.13
Surface temperature (°C)	8537	16 930	5150	2795

   a Describe the relationship between mass and surface temperature.

   b Suggest a reason why there is this relationship.

   c Estimate the surface temperature of the Sun to the nearest 1000 °C.

Life cycles of stars

Learning objective:

- describe the life cycles of a star like the Sun and a massive star.

When our Sun becomes a red giant, in 5 billion years time, it will expand to somewhere between the orbits of Earth and Venus. It has been predicted that the Earth will then be swallowed up by the Sun and destroyed.

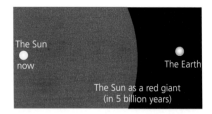

The Sun now

The Earth

The Sun as a red giant (in 5 billion years)

Figure 8.18 The sizes of the Sun and the Earth compared with a red giant

Life cycle of a star like the Sun

What happens to a star after all the hydrogen has been used up depends on its mass.

In a small star like the Sun, when all of the hydrogen has fused to form helium, it becomes unstable – it is no longer a main sequence star. The outer layers of the star are pushed out to many times the stars' original size. The star has expanded to become a **red giant** (Figure 8.19).

Helium starts fusing to form other elements. When the helium has been used up, nuclear fusion stops. The star stops emitting energy and the outer layers of the star expand to become a cloud of ionised gas called a planetary nebula (Figure 8.20).

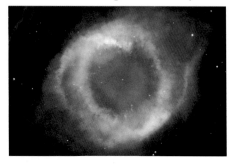

Figure 8.20 The Helix planetary nebula

As the core continues to cool and dim it becomes a **white dwarf**. When it stops shining, the now-dead star is called a **black dwarf**.

1. What does the life cycle of a star depend on?
2. When does a star leave the main sequence of its life?
3. Describe how a star like the Sun changes from being a red giant.

The life cycle of a massive star

Massive stars can be between three and 50 times more massive than our Sun. These stars have a much shorter life span: only millions of years, not billions. This is because they use up their hydrogen at a much faster rate.

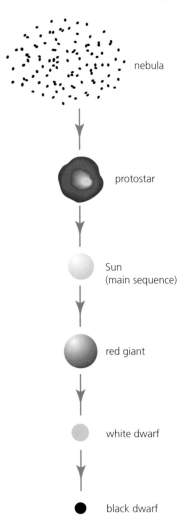

nebula

protostar

Sun (main sequence)

red giant

white dwarf

black dwarf

Figure 8.19 The stages in the life cycle of a star like the Sun

When the hydrogen is used up, the star expands to become a **red super giant**, and helium nuclei fuse to form larger nuclei of elements such as carbon and nitrogen.

For the next few million years the red super giant contracts and expands. With each cycle the temperature increases and more of the lighter nuclei fuse together.

4 Why do massive stars have such a short lifespan compared with smaller stars like our Sun?

Neutron stars and black holes

The star finally reaches a point where fusion stops. The core collapses causing a gigantic explosion called a **supernova** (Figure 8.21) which blows the outer layers of the star into space. The supernova distributes the elements formed by the fusion processes throughout the Universe.

Figure 8.22 The Crab Nebula is the remnants of a supernova

If the surviving core is between 1.5 and 3 solar masses it contracts to become a tiny, very dense **neutron star**. If the core is much greater than 3 solar masses, the core contracts to become a **black hole**. There is a neutron star at the centre of the Crab Nebula (Figure 8.22).

5 Describe what happens to a red super giant.

6 Suggest how the lifecycle of stars continues after a supernova.

7 The stars in the night sky don't appear to be changing. Suggest why we know they will pass through different stages depending on their mass.

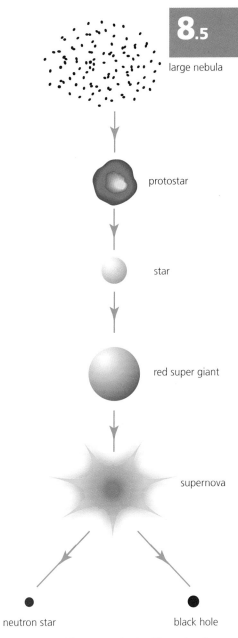

large nebula

protostar

star

red super giant

supernova

neutron star   black hole

Figure 8.21 The stages in the life cycle of massive stars

DID YOU KNOW?

The explosion that created the Crab Nebula was observed by Chinese astronomers in 1054 and was visible in daylight. It was observed for over 600 days.

How the elements are formed

Learning objectives:

- understand how new elements are produced by nuclear fusion inside a star
- recognise that the heavier elements are made in a supernova.

KEY WORDS

nuclear fusion
supernova

Every atom that we are made of – the carbon in our bones, the calcium in our teeth, the nitrogen in our DNA – was created by fusion in a star or in a supernova.

The early Universe

The early Universe contained only clouds of hydrogen with a little helium. The clouds of gas were compressed by gravity to form the first stars. The hydrogen nuclei began to fuse together and form helium (Figure 8.23).

DID YOU KNOW?

It takes thousands and possibly millions of years for energy to reach the surface of the Sun from its core. But then it only takes eight minutes for the energy to reach us on Earth.

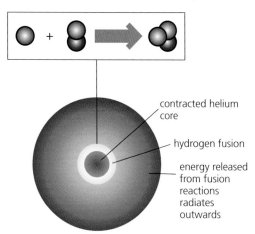

Figure 8.23 Fusion in a star like the Sun

1. The early Universe was very hot and dense. Suggest why there was some helium within the clouds of hydrogen.

2. What was the effect of gravity?

3. a Where does fusion take place in a star like the Sun?

   b Explain why fusion doesn't happen in the other parts of the Sun.

Making light elements

All elements with atomic masses up to iron are made by **nuclear fusion** inside a star. The more massive the star, the greater the range of elements that are created inside it.

When the hydrogen inside a star starts to run out, the core nuclear fusion reaction begins to slow down and the force of gravity causes the star to contract. As the star contracts, it compresses the core and its temperature rises, until the temperature at which helium fuses is reached. This starts a new nuclear fusion reaction, which produces larger nuclei of heavier elements – first beryllium, and then carbon.

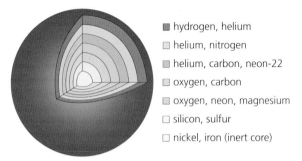

- ■ hydrogen, helium
- □ helium, nitrogen
- ■ helium, carbon, neon-22
- □ oxygen, carbon
- □ oxygen, neon, magnesium
- □ silicon, sulfur
- □ nickel, iron (inert core)

Figure 8.24 Elements created inside a red super giant

The expansion and contraction cycle continues, forming oxygen, sodium, magnesium, right up to nickel and iron. When it reaches iron, it cannot produce larger elements because energy is not released when the nuclei fuse. Energy is needed to make iron nuclei fuse into larger, heavier elements.

4 What happens to a star when its hydrogen fuel is used up?

5 Explain how elements lighter than iron are produced inside a star.

6 Why is iron the largest element that can be fused in a red super giant?

Making elements heavier than iron

The red super giant has a core of nickel and iron surrounded by layers of the lighter elements. When fusion stops no more energy is released. Gravity causes the star to collapse in on itself and explode in a supernova.

In a **supernova**, the temperature becomes high enough for the heavier elements such as uranium and gold to be formed. The force of the explosion scatters all these heavier elements and the elements formed by fusion processes out into space to join with other gas and dust clouds to become a new nebula. This becomes the nursery for the next generation of star formation.

7 Explain why the heavier elements can only be created in a supernova.

8 Suggest why elements heavier than iron can be found inside the Sun.

DID YOU KNOW?

With each contraction and expansion of a red super giant and increase in temperature a new element is created.

DID YOU KNOW?

Our Sun is about 4.6 billion years old. The Universe is estimated to be 13.7 billion years old. So there have been complete life cycles of stars before our Sun was born.

Red-shift

Learning objectives:

- describe red-shift
- describe evidence for the expanding Universe
- recognise that that there is still much about the universe that is not understood, for example dark mass and dark energy.

KEY WORDS

Big Bang
dark energy
dark matter
red-shift
wavelength

Red-shift is used in astronomy to help determine the age of the Universe.

Red-shift

If an object was emitting light (or any other electromagnetic wave) and travelling towards you very quickly, this would have the effect of compressing the waves. The distance between them would be less. If the light source was travelling away from you very quickly, this would have the effect of stretching the waves. The distance between them would be greater.

When the light source is moving away from you, the waves appear to stretch and their wavelength becomes longer so the frequency decreases. The light becomes redder.

KEY INFORMATION

The greater the speed of the light source relative to the observer, the greater the change in wavelength of the light observed.

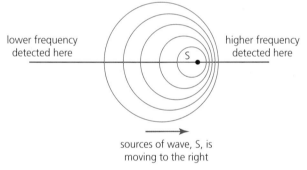

Figure 8.25 The wavelengths and frequencies of electromagnetic waves detected from a source depend on how the source is moving relative to the observer

KEY INFORMATION

In the early part of the 20th century, astronomers thought that the Universe was not static but dynamic. Hubble was able to prove it with his direct observations of the red-shifted spectral lines of galaxies.

From around 1912, astronomers including Edwin Hubble noticed that the light from distant galaxies had longer wavelengths than expected. The light had **wavelengths** shifted towards the red end of the spectrum (Figure 8.26). They named this **red-shift**.

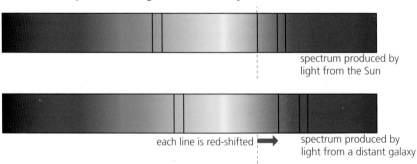

Figure 8.26 The dark lines in the spectrum from a star are caused when atoms in a gas like helium absorb light of particular wavelengths. The whole pattern of lines from a distant galaxy has been red-shifted compared with light from the Sun

1 Explain what red-shift means.

2 What does red-shift tell us about galaxies?

Evidence for an expanding Universe

Everywhere Hubble looked, the galaxies were nearly always moving away from us. This was the first observed evidence that supported the theory that the Universe was expanding.

He estimated that the degree of red-shift was directly proportional to the distance of the galaxy from the Earth. This meant that the further away a galaxy, the bigger the red-shift and hence the faster it is moving away from us. Hubble concluded that the Universe is expanding. Most galaxies will see other galaxies moving away from them (Figure 8.27). The further away a galaxy is from another galaxy, the faster it is moving away from it.

The expansion of the Universe suggests that in the past it was much smaller and more dense than it is today. There must have been a point at which it started to expand, which we call the **Big Bang**. The Big Bang theory also suggests that this small dense region was extremely hot.

Red-shift and other observations such as from radio telescopes made it possible to develop and support theories such as the Big Bang theory.

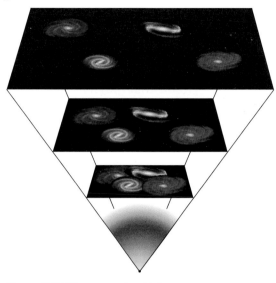

Figure 8.27 The expanding Universe

3. **What relationship did Hubble discover between the distance of a galaxy from the Earth and its speed?**

4. **Explain what this evidence suggests about our Universe.**

5. **Suggest why observations are useful in developing theories.**

6. **Describe the Big Bang theory.**

7. **If all the galaxies are moving away from us, are we at the centre of the Universe?**

DID YOU KNOW?

If the red-shift of galaxies changed to blue-shift, we would know that the Universe had stopped expanding and begun contracting back in on itself.

New questions about the Universe

When astronomers calculated the mass of galaxies they found something very odd. The mass of the visible stars is not enough to hold the stars in orbit around the galaxy. There must be extra matter that we cannot see, which holds galaxies together by gravitational attraction. We call this matter **dark matter**.

Astronomers have also discovered from observations of supernovae that the rate of expansion of the Universe appears to be increasing. For this to be happening, most of the mass/energy of the Universe must consist of something we know nothing about, which is called **dark energy**. Scientists do not yet understand either dark matter or dark energy.

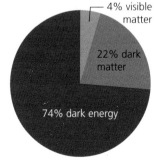

Figure 8.28 Percentages of visible matter, dark matter and dark energy in the Universe

8. **What surprised astronomers when they tried to calculate the mass of the Universe?**

KEY CONCEPT

Gravity: the force that binds the Universe

KEY WORD

gravitational field strength

Learning objectives:

- understand that gravity provides the force that keeps planets and satellites in orbits
- understand that gravity is necessary at the start of a star's life cycle and to maintain equilibrium in a stable star
- describe how the weight of an object depends on the gravitational field strength.

Every time you drop a book, step onto the bathroom scales or throw a ball into the air you experience the force of gravity. Gravity is responsible for the shape, complexity and dynamics of the Universe but there is still much about the Universe that is not understood.

What is gravity?

Gravity is a force that attracts all things with mass towards one another. It is responsible for the formation of stars and galaxies and holds planets in orbit around their stars. Even the Milky Way is held in orbit by the common force of gravity of all of its matter (including a black hole at the centre of the galaxy).

Gravity is a non-contact force, and has an infinite range. The force just becomes weaker with distance from the object (Figure 8.30). Gravity always pulls objects towards the centre of the Earth (Figure 8.31).

Gravity holds our atmosphere wrapped around the Earth. Without gravity there would be no force to hold the planets in orbit around the Sun or to keep the Moon in orbit around the Earth. Without gravity the stars would never have come into existence and our Universe would have just remained a cloud of gas.

DID YOU KNOW?

The force of gravity actually distorts space and time. This means it bends light and it makes clocks run more slowly.

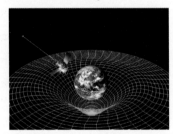

Figure 8.29 Spacetime distortion generated by the mass of an object

1. Which of these objects can exert a gravitational force?

   a a star   b a planet
   c a mouse   d an atom

2. What is the role of gravity in the Universe?

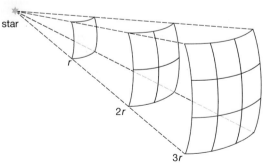

star

r

2r

3r

Figure 8.30 The force of gravity gets weaker as distance increases because it acts over a larger area

Gravitational field strength

The weight of an object is the force acting on it due to gravity. The unit of force is the newton (N). Here on Earth each kilogram is pulled down with a force of 9.81 N. We say the **gravitational field strength** is 9.81 N/kg (newtons per kilogram). This is given the symbol g.

We can also refer to the force of gravity in terms of the acceleration this force gives to a falling object. This is called the acceleration due to gravity and has a unit of m/s^2 (metres per second squared). Here on Earth the numerical value is $9.81\ m/s^2$. This is the same value as g, just with different units. From Newton's second law of motion, $F = ma$ (force = mass x acceleration), the acceleration is given by $a = F/m$. This means the unit of acceleration (metres per second squared) is equivalent to a unit of gravitational field strength (newtons per kilogram).

On Mars, the gravitational field strength is lower – about $3.7\ m/s^2$ (Figure 8.32). This means that the weight of an object on Mars is less than the weight of the same object on Earth.

Figure 8.31 The force of gravity pulls us towards the centre of the Earth wherever we are on the Earth

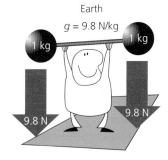

Earth
g = 9.8 N/kg

1 kg   1 kg

9.8 N   9.8 N

3 How does gravity affect our weight?

4 An elephant has a mass of 1000 kg.

    a Calculate the weight of the elephant on Earth.

    b Use $F = ma$ to determine the acceleration of the elephant if it was falling to the ground.

    c Would you get the same answer for (b) for an object of a different mass?

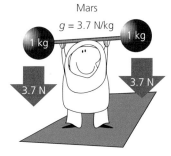

Mars
g = 3.7 N/kg

1 kg   1 kg

3.7 N   3.7 N

Figure 8.32 Things weigh less on Mars because the force of gravity is weaker for an object of the same mass

MATHS SKILLS

Using scale and standard form

Objectives:

- understand the scale of objects in the Universe
- use standard form.

KEY WORDS

scale
standard form

When looking at the scale of the Universe, we need a different way of expressing the numbers involved – it becomes very difficult to understand the size of a number when there are lots of zeros in it.

Scale

The Moon is one quarter the diameter of the Earth. That's a ratio of 1:4. The Moon is the largest natural satellite of any planet in the Solar System.

Mars is only about half the diameter of the Earth: a linear ratio of 1:2.

The diameter of Jupiter, the largest planet, is eleven times that of the Earth, and its volume is over 1300 times greater than the Earth.

The Sun's diameter is 109 times greater than the Earth and 1 300 000 Earths could fit inside the Sun.

When you draw a **scale** model of the Solar System, there are problems with sizes. If you made the Sun the size of a large beach ball, the Earth would be the size of a pea and stand 100 m from the beach ball. Jupiter would be the size of a tennis ball and 520 m from the beach ball. Neptune, the furthest planet from the Sun, would be 3 km away.

If you want to represent the Solar System on a standard A4 sheet of graph paper, you need to take the longest distance, in this case to Neptune, and scale it down to fit the longest side of the paper.

1. **Look at the data set in topic 8.2.**

   Draw a scale model of the distances of the planets to the Sun to fit on an A4 piece of paper.

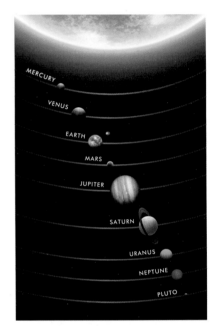

Figure 8.33 The Sun and planets (including the dwarf planet Pluto), not to scale

DID YOU KNOW?

There is a scale model of the Solar System on the south side of York. The sizes of the Sun and planets and the distances between them are all to the same scale. The model is about 10 km long! You can walk between the Sun and planets.

There is also a scale model at Anstruther in Scotland.

Standard form

The distance from the Sun to the Earth is about 150 000 000 km. The distance to Neptune is 4500 million km. The distance to Alpha Centauri is 41 000 000 000 000 km. And our Milky Way galaxy is 1 000 000 000 000 000 000 km in diameter. We handle such large numbers by expressing them in **standard form**.

A number in standard form is written as:

$A \times 10^B$

where A is a decimal number between 1 and 10 (including 1 but not 10), and B is an integer.

Here are some examples:

1000 km = 1×10^3 km

This is because $1000 = 1 \times 10 \times 10 \times 10$.

Distance to the Sun, 150 000 000 km = 1.5×10^8 km

To write 150 000 000 km in standard form, imagine moving the first digit on the left, 1, to the right, until the decimal point would be between the 1 and 0. Count how many places you have to move this digit. We have to move the 1 eight places to the right to reach the decimal point. So the answer is 1.5×10^8 km.

> **KEY INFORMATION**
>
> Writing the distance to the Sun as 15.0×10^7 km is correct, but it is not in standard form because the first number is greater than 10.

2 Write the distances from the Sun to Neptune, to Alpha Centauri, and the diameter of the Milky Way in standard form.

Another way of handling such large numbers is to use a larger unit. Astronomers use the light year, which is the distance light will travel in a year of 365 days.

3 a Calculate how far will light travel in a year. (Speed of light is 3.0×10^8 m/s.)

b Use your answer to calculate the diameter of the Milky Way in light years.

Negative powers of ten for very small numbers

Very small numbers can also be written in standard form. For example, the radius of an atom is about 1×10^{-10} m. A negative power of ten means a number less than 1. In standard form, 1/1 000 is 1×10^{-3}, 1/1 000 000 is 1×10^{-6} and 1/1 000 000 000 is 1×10^{-9}. These correspond to the SI prefixes milli, micro and nano.

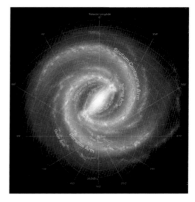

Figure 8.34 Artist's concept of the Milky Way.

4 The change in wavelength (red-shift) for light from a receding galaxy is 170 nm. Write this in metres, in standard form.

5 Here are some approximate volumes:

proton = 1×10^{-45} m³

person = 1 m³

Solar System = 1×10^{39} m³

a Calculate how many times bigger a person is, compared with a proton inside them.

b How much bigger is the Solar System compared with a person?

c If a proton was scaled up to the size of a person, and the person was scaled up by the same factor, how many Solar Systems would the volume of the person be?

Check your progress

You should be able to:

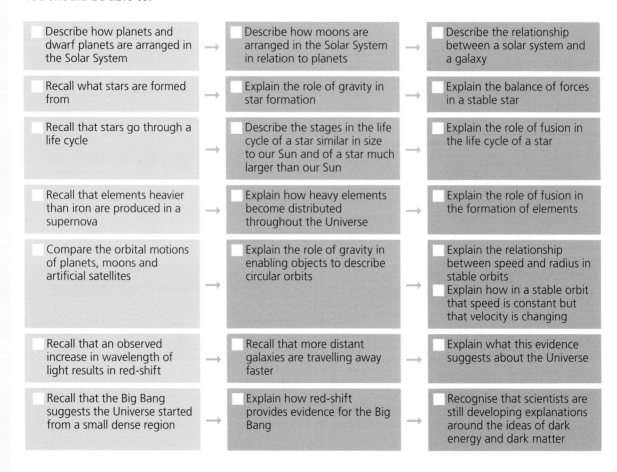

Describe how planets and dwarf planets are arranged in the Solar System → Describe how moons are arranged in the Solar System in relation to planets → Describe the relationship between a solar system and a galaxy

Recall what stars are formed from → Explain the role of gravity in star formation → Explain the balance of forces in a stable star

Recall that stars go through a life cycle → Describe the stages in the life cycle of a star similar in size to our Sun and of a star much larger than our Sun → Explain the role of fusion in the life cycle of a star

Recall that elements heavier than iron are produced in a supernova → Explain how heavy elements become distributed throughout the Universe → Explain the role of fusion in the formation of elements

Compare the orbital motions of planets, moons and artificial satellites → Explain the role of gravity in enabling objects to describe circular orbits → Explain the relationship between speed and radius in stable orbits
Explain how in a stable orbit that speed is constant but that velocity is changing

Recall that an observed increase in wavelength of light results in red-shift → Recall that more distant galaxies are travelling away faster → Explain what this evidence suggests about the Universe

Recall that the Big Bang suggests the Universe started from a small dense region → Explain how red-shift provides evidence for the Big Bang → Recognise that scientists are still developing explanations around the ideas of dark energy and dark matter

Worked example

1 **Describe the orbit of an artificial satellite.**

It goes round the Earth.

Yes satellites go round the Earth but the answer could be improved by describing what kind of orbit, i.e. circular.

2 **In what way do the observations made by Hubble support the Big Bang?**

The observation shows there is red-shift.

Just saying red-shift is not really enough. The answer should go into more detail to explain what is meant by red-shift, what Hubble observed, what it implied and why he called this effect red-shift.

3 **Explain the differences and similarities between moons and artificial satellites.**

They both orbit planets, but one of them has been put into orbit.

This is correct but does not give enough information. You should aim to state what the similarities are, and what the differences are – and which object the differences apply to, e.g. moons are natural satellites, but artificial satellites have been put into orbit by humans.

4 **What does it mean when a star is in its stable period?**

It's in the main sequence.

Stable period and main sequence are different names for the same thing.

You need to show that you understand and describe what is happening inside the star when it is in its stable period.

5 **Every star goes through a cycle of birth, life and death. Explain how the mass of a star determines its life cycle.**

A small star has a long life cycle and becomes a white dwarf. A big star has a much shorter life cycle and blows up.

The answers needs to be more precise with the words used. Yes it has to do with size, but in what way?

You should describe what happens to an average sized star and then describe the stages that a much more massive star goes through. Structure your answer carefully and use the correct scientific terms.

End of chapter questions

Getting started

1. State what is at the centre of our Solar System. **1 Mark**

2. State the name of the force that pulls together the dust and gas of a nebula to begin forming a new star. **1 Mark**

3. What is the Milky Way? **1 Mark**

   A a star   B a solar system   C a galaxy   D a globular cluster

4. Name the process that happens when two small nuclei join together to form a larger nucleus. **1 Mark**

5. Describe what will happen to the Sun in the future. **2 Marks**

6. If a satellite is to be moved to a lower orbit around the Earth, how does its speed have to alter? **1 Mark**

7. Exoplanets are planets that orbit stars other than the Sun. Here is some data for two exoplanets that orbit the same star.

	Kepler 33-b	Kepler 33-f
Time for a complete orbit	5.6 days	41 days

   Which planet orbits closest to the star? **1 Mark**

8. Explain why do we not see blue-shift when we observe galaxies. **2 Marks**

Going further

9. Name the planets of our Solar System in order of distance from the Sun. **1 Marks**

10. Which stage of a star's life takes the longest time? **1 Marks**

    A protostar   B main sequence   C red giant   D supernova

11. Explain how a protostar is formed. **2 Marks**

12. Here is some data about the planets in the Solar System.

Planet	Mass ($\times 10^{24}$ kg)	Diameter ($\times 10^{7}$ m)	Gravitational field strength at the surface (N/kg)
Mercury	0.33	0.49	3.7
Venus	4.87	1.21	8.9
Earth	5.97	1.28	9.8
Mars	0.64	0.68	3.7
Jupiter	1898	14.30	23.1
Saturn	568	12.05	9.0
Uranus	86.8	5.11	8.7
Neptune	102	4.95	11.0

  a Name the planet that has the largest gravitational field strength and suggest a reason for this.

`2 Marks`

  b An object weighs 111 N on the surface of Mars. Determine how much it would weigh on the surface of the Earth.

`2 Marks`

13 A TV satellite dish is fixed. Explain what kind of orbit the satellite that transmits the signals is in.

`2 Marks`

More challenging

14 Explain what is meant by red-shift.

`1 Mark`

15 Name one use for an artificial satellite.

`1 Mark`

16 Explain what happens when a star becomes a red giant.

`2 Marks`

17 Here is some data for three stars that are currently in the main sequence stage.

	Proxima Centauri	Tau Ceti	Zeta Ophiuci
Mass compared to the mass of the Sun (mass of the Sun = 1)	0.12	0.78	20

Describe how these stars will evolve in the future.

`6 Marks`

Most demanding

18 Describe what is happening inside a star when it is at the main sequence stage.

`2 Marks`

19 A star has a core of nickel and iron. Describe what will happen to the star.

`2 Marks`

20 The graph below shows data similar to the data Hubble used in 1929 to propose the Big Bang theory.

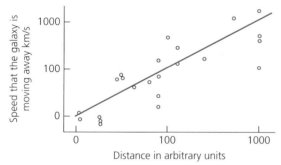

Explain the Big Bang theory and evaluate the data that Hubble used in 1929.

`6 Marks`

`Total: 40 Marks`

Appendix – Circuits

 switch (open)

 lamp

 switch (closed)

 fuse

 cell

 voltmeter

 battery

 ammeter

 diode

 thermistor

 resistor

 variable resistor

 LDR

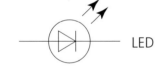 LED

Appendix – Equations

It is important to be able to recall and apply the following equations using standard units:

Equation Number	Word Equation	Symbol Equation
1	weight = mass × gravitational field strength	$W = mg$
2	work done = force × distance (along the line of action of the force)	$W = Fs$
3	force applied to a spring = spring constant × extension	$F = ke$
4	moment of a force = force × distance (normal to direction of force)	$M = Fd$
5	pressure = $\dfrac{\text{force normal to a surface}}{\text{area of that surface}}$	$p = \dfrac{F}{A}$
6	distance travelled = speed × time	$s = vt$
7	acceleration = $\dfrac{\text{change in velocity}}{\text{time taken}}$	$a = \dfrac{\Delta v}{t}$
8	resultant force = mass × acceleration	$F = ma$
9	momentum = mass × velocity	$p = mv$
10	kinetic energy = 0.5 × mass × (speed)^2	$E_k = \frac{1}{2}mv^2$
11	gravitational potential energy = mass × gravitational field strength × height	$E_p = mgh$
12	power = $\dfrac{\text{energy transferred}}{\text{time}}$	$P = \dfrac{E}{t}$
13	power = $\dfrac{\text{work done}}{\text{time}}$	$P = \dfrac{W}{t}$
14	efficiency = $\dfrac{\text{useful output energy transfer}}{\text{useful input energy transfer}}$	
15	efficiency = $\dfrac{\text{useful power output}}{\text{total power output}}$	
16	wave speed = frequency × wavelength	$v = f\lambda$
17	charge flow = current × time	$Q = It$
18	potential difference = current × resistance	$V = IR$
19	power = potential difference × current	$P = VI$
20	power = (current)^2 × resistance	$P = I^2R$

Equation Number	Word Equation	Symbol Equation
21	energy transferred = power × time	$E = Pt$
22	energy transferred = charge flow × potential difference	$E = QV$
23	density = $\dfrac{\text{mass}}{\text{volume}}$	$\rho = \dfrac{m}{V}$

The following equations will appear on the equations sheet that you are given in the exam. It is important to be able to select and apply the appropriate equation to answer a question correctly.

Equation Number	Word Equation	Symbol Equation
1	pressure due to a column of liquid = height of column × density of liquid × gravitational field strength	$p = h\rho g$
2	(final velocity)² − (initial velocity)² = 2 × acceleration × distance	$v^2 - u^2 = 2as$
3	force = $\dfrac{\text{change in momentum}}{\text{time taken}}$	$F = \dfrac{m\Delta v}{\Delta t}$
4	elastic potential energy = 0.5 × spring constant × (extension)²	$E_e = \frac{1}{2}ke^2$
5	change in thermal energy = mass × specific heat capacity × temperature change	$\Delta E = mc\Delta\theta$
6	magnification = $\dfrac{\text{image height}}{\text{object height}}$	
7	period = $\dfrac{1}{\text{frequency}}$	
8	force on a conductor (at right-angles to a magnetic field) = magnetic flux density × current × length	$F = BIl$
9	thermal energy for a change of state = mass × specific latent heat	$E = mL$
10	$\dfrac{\text{potential difference across primary coil}}{\text{potential difference across secondary coil}} = \dfrac{\text{number of turns in primary coil}}{\text{number of turns in secondary coil}}$	$\dfrac{V_p}{V_s} = \dfrac{n_p}{n_s}$
11	potential difference across primary coil × current in primary coil = potential difference across secondary coil × current in secondary coil	$V_s I_s = V_p I_p$
12	For gases: pressure × volume = constant	$pV = \text{constant}$

Glossary

A

absorption process in which matter takes in energy, e.g. when an atom takes in energy from an electromagnetic wave

acceleration rate at which an object speeds up, calculated from change in velocity divided by time; symbol a, unit metres per second squared, m/s^2

accuracy how close a measurement is to its true value

activity the rate at which unstable nuclei decay in a sample of a radioactive material; unit becquerel, Bq

air resistance force produced by the collision of molecules in air with a moving object; the force acts to oppose the direction of movement

alternating current (a.c.) electric current that continually changes direction

alpha particle a particle (two neutrons and two protons, same as a helium nucleus) emitted by an atomic nucleus during radioactive decay

alternator device that produces an alternating current using the generator effect, e.g., from the rotation of a coil in a magnetic field

ammeter device that measures electric current

ampere SI unit of electric current, symbol A; an electric current of 1 A is equal to 1 C of charge passing through a point in a circuit in 1 s

amplitude maximum displacement of a wave or oscillating object from its rest position

angle of incidence the angle between the incident ray and the normal

angle of reflection the angle between the reflected ray and the normal

atmospheric pressure force per unit area produced by the weight of air; it decreases as you go higher in the atmosphere

atomic number number of protons in the nucleus of an atom of an element

attraction force that acts between two objects that tends to draw them closer together, e.g., the force that arises between a positive electric charge and a negative electric charge

average speed distance travelled by an object over a measured time interval

B

background radiation nuclear radiation that is present in the environment as a result of the radioactive decay of certain elements; it is produced from natural sources (e.g., from radon gas in the Earth's crust) and from artificial sources (e.g., as a result of testing nuclear weapons)

becquerel unit of activity for a radioactive isotope, symbol Bq

beta particle fast-moving electron that is emitted by an atomic nucleus in some types of radioactive decay

Big Bang theory generally accepted explanation for the origin of the universe, which states that the universe began from a very small region that was extremely hot and dense

black body object that absorbs all the radiation that falls on it

black dwarf star at the end of its life, where no further nuclear fusion reactions take place and so no light is produced

boundary line or surface between two different media (e.g., the surface between air and glass), across which light may be refracted or reflected

braking distance distance travelled by a vehicle after the brakes have been applied before coming to a complete stop

C

cancer disease in which uncontrolled division of cells takes place, forming a tumour; some types of cancer are caused by the effects of ionising radiation on cells

cell (electric circuits) circuit component that stores energy in the form of chemical energy a number of electric cells may be joined to form a battery

cell (living things) smallest structural unit of a living organism; cells may be damaged by some types of ionising radiation

centre of mass a single point where the weight of an object can be considered to act

chain reaction occurs when the neutrons released by a nucleus undergoing fission go on to split more nuclei

change of state process in which one state of matter changes to another, e.g., solid melting to form liquid, or gas condensing to form liquid

circuit diagram visual representation of electrical components connected by wires, using standard symbols

circuit symbol visual representation of an electrical component; a standard symbol exists for each type of electrical component

closed system system that is not acted upon by any external forces and does not exchange energy with its surroundings

commutator (split-ring) electrical connector consisting of two semi-circular metal contacts, which is connected to a rotating coil in a dynamo (to produce direct current) or a motor (to use direct current)

compass (magnetic) device that shows the direction of a magnetic field in which it is placed; typically a compass is made from a magnetised needle or pointer that is free to rotate

components (forces) the result of breaking down or resolving a single force into two separate forces acting in perpendicular directions, often horizontally and vertically

compress squashing something into a smaller volume – liquids are incompressible: they cannot be squashed

compression a region of a sound wave where the particles are closer together

concave lens lens that curves inwards with the centre thinner than the edges – it causes light rays to spread out (diverge). The image produced by a concave lens is always virtual

conclusion section at the end of a report on an experiment that summarises how the results support or contradict the original hypothesis

conductivity (thermal) quantity that measures the rate at which thermal energy is conducted through a material; the higher the thermal conductivity, the higher the rate of thermal energy transfer

conductor material or object that allows electric charge to flow through it

conservation of electric charge a fundamental principle of physics: electric charge cannot be created or destroyed, only transferred from one object to another; the total electric charge of a closed system is constant

conservation of energy a fundamental principle of physics: energy cannot be created or destroyed, only transferred, stored or dissipated. This means that the total energy of a closed system is constant.

conservation of mass a fundamental principle of physics: mass cannot be created or destroyed by physical changes or chemical reactions; the total mass of a system is constant

conservation of momentum a fundamental principle of physics: the total momentum of a system of objects after a collision is the same as the total momentum before the collision

contact force force that acts at the point of contact between two objects (e.g., friction or the normal reaction)

contamination (radioactivity) the unwanted presence of materials containing radioactive atoms

control variable quantity in an experiment that is kept constant while the independent variable is changed and the dependent variable is measured

convex lens lens that curves outwards with the centre thicker than the edges – it causes light rays to come together (converge). The image produced by a convex lens can be either real or virtual

coulomb SI unit of electric charge, symbol C

count rate the number of decays recorded each second by a detector (e.g. Geiger–Muller tube)

CT (or CAT) scan scan that takes many X-ray images from different directions to build up a 3D image of the body

current (electric) flow of electric charge; the size of the electric current is the rate of flow of electric charge; symbol I, unit amps (A)

D

dark energy unknown form of energy which is thought to accelerate the expansion of the universe. It is estimated to constitute around 68% of the total energy in the universe

dark matter unidentified matter which doesn't reflect electromagnetic radiation and is thought to constitute around 27% of the total mass-energy of the universe

data measurements of quantities in an experiment; data can be recorded in a table and used to produce graphs

decay (radioactive) process in which the nucleus of an atom breaks apart and emits radiation

deceleration negative acceleration, when an object slows down

degrees Celsius unit of temperature, symbol °C; 1 °C is $\frac{1}{100}$ of the temperature difference between the melting point and boiling point of water

density measure of the amount of substance per unit volume; symbol ρ, unit kilograms per metre cubed, kg/m^3

dependent variable quantity in an experiment that is measured for each change in the independent variable

diffuse reflection reflection from a rough surface, so that an incident ray is scattered (reflected at many angles)

diode circuit component with a low resistance if it is connected one way around in a circuit, or a very high resistance if it is connected the other way around; a diode allows current in one direction only

direct current (d.c.) flow of electric charge in one direction only

directly proportional relationship a relationship of the form y proportional to ($\propto$) x; plotting a graph of y against x will produce a straight line through the origin

displacement distance moved in a particular direction; it is a vector quantity and is equal to the area under a velocity–time graph; symbol s, unit metres, m

dissipation the spreading out of energy into the environment, so that it is stored in less useful ways

distance quantity that measures how far an object moves but not the direction; it is a scalar quantity; symbol s, unit metres, m

distance–time graph graph with distance on the y-axis and time on the x-axis; the gradient of a distance–time graph is equal to the speed

dose (radiation) quantity that measures the amount of ionising radiation received (e.g., by a human body); measured in sieverts (Sv)

dynamo device that produces a direct current using the generator effect (e.g., from the rotation of a coil in a magnetic field, using a commutator)

E

earth (electrical) electrical connection between the metal case of an electrical appliance and the ground, used as a safety device to prevent the case becoming charged if touched by a live wire

echo reflection of a sound or ultrasound wave

echo sounding process in which high-frequency sound waves are reflected off a surface to measure distances (e.g., a ship measuring the depth of the seabed)

efficiency useful output energy transfer divided by the total input energy transfer (or useful power output divided by the total power input) – may be expressed as a percentage or as a decimal

elastic deformation process in which an object (e.g., a spring) is stretched by a force and returns to its original size or length when the force is removed

elastic potential energy energy that is stored in an object as a result of the object being stretched or compressed

electric charge fundamental property of matter that results in an electric field around an object; objects can be electrically positive, negative or neutral (no overall charge); symbol Q, unit coulomb, C

electric current flow of electric charge; the size of the electric current is the rate of flow of electric charge; symbol I, unit amps, A

electric field region around an electrically charged object in which a force acts on other electrically charged objects; the objects do not have to touch for the force to act

electrical power amount of energy transferred each second; symbol P, unit watts or joules per second, W or J/s

electromagnet magnet formed by an electric current flowing through a solenoid (coil of wire) with an iron core – the magnetic field of an electromagnet can be switched on and off by switching the current on and off

electromagnetic (EM) spectrum electromagnetic waves ordered according to wavelength and frequency – ranging from radio waves to gamma rays

electron a particle that orbits the nucleus in all atoms; electrons have a negative electric charge

emission process in which energy or a particle are given out by an atom (e.g., in beta particle emission, a high-speed electron is given out)

energy level stable state of a physical system, e.g., electrons orbiting the nucleus of an atom can exist in only particular energy levels and be moved between them

energy store the energy associated with a fuel (chemical energy store), heated object (thermal energy store), moving object (kinetic energy store), a stretched spring (elastic potential energy store) and an object raised above ground level (gravitational potential energy store)

energy transfer process in which energy is moved from one store to another

equilibrium situation in which there is no resultant force acting on an object; any forces that are acting balance each other in magnitude and direction

estimate best guess of the value of a quantity that cannot be measured precisely, based on scientific knowledge and observation

extension the increase in length of an object when a force is applied

F

filter (optical) thin coloured sheet of material that allows only particular wavelengths of light through it

fission (nuclear) process in which the nucleus of an atom splits to form nuclei of atoms with lower atomic numbers

Fleming's left-hand rule shows which way a current-carrying wire tries to move when placed in a magnetic field – First finger shows Field, seCond finger shows Current, thuMb shows Movement

fluid substance that is a liquid or gas

focal length distance between the centre point of a lens and the point (the principal focus) where parallel rays entering the lens either cross over or appear to have come from

force 'push' or 'pull' on an object that can cause the object to accelerate

free body (force) diagram drawing showing the magnitude and direction of all the forces acting on an object, and in which the object is represented by a point

frequency number of waves passing a set point in one second

friction force acting at points of contact between objects moving over each other, to resist the movement

fusion (nuclear) process in which two nuclei of different atoms merge to form a single nucleus of one atom with a larger atomic number

G

gamma radiation ionising electromagnetic radiation with shortest wavelengths in the electromagnetic spectrum

gas state of matter in which all the particles of a substance are separate and move about freely

generator effect when a conducting wire is moved so that it cuts across magnetic field lines; a potential difference is induced across the ends of the wire

global warming increase in the mean surface temperature on Earth

gradient the slope of a graph = $\dfrac{\text{change in } y\text{-axis values}}{\text{change in } x\text{-axis values}}$

gravitational field strength quantity that measures the 'pull' of the force of gravity on each kilogram of mass; symbol g, unit newtons per kilogram, N/kg

gravitational potential energy energy that an object has because of its position; e.g., increasing the height of an object above the ground increases its gravitational potential energy

gravity, force due to attractive force acting between all objects with mass; gravity pulls objects downwards on Earth and keeps planets and satellites in orbit

H

half-life the average time it takes for half of all the nuclei present in a sample of a radioactive element to decay, or the time it takes for the count rate (or activity) to halve

hazard anything that may cause injury (e.g., the risk of contamination by radioactive materials when they are used in a scientific experiment)

headphones very small loudspeakers designed to produce sound waves from an electrical signal and cause the air inside the ear to vibrate

hertz a unit of frequency, equivalent to one cycle per second

hypothesis an idea for, or explanation of, a phenomenon in science that can be tested through study and experimentation

I

incident ray a ray of light that strikes a surface or boundary between two media

independent variable quantity in an experiment that is changed or selected by the experimenter

induced current electric current that arises in a conducting wire placed in a magnetic field when the wire is part of a complete circuit and is moved so that it cuts across magnetic field lines

induced magnet material that is magnetic only when it is placed in the magnetic field of another magnet, e.g., the iron core within a solenoid

induced potential difference potential difference that arises between the ends of a conducting wire placed in a magnetic field when the wire is moved so that it cuts across magnetic field lines

inelastic deformation process in which an object (e.g., a spring) is stretched by a force and permanently deformed so that it does not return to its original size or length when the force is removed

inertia natural tendency of objects to resist changes in their velocity

inertial mass measure of how difficult it is to change the velocity of an object

infrared radiation electromagnetic radiation with a range of wavelengths longer than visible light but shorter than microwaves; emitted in particular by heated objects

insulator (electrical) material that does not allow electric charges to pass through it (e.g., wood)

insulator (thermal) material that does not easily allow the transfer of thermal energy

intercept point at which the line of a graph crosses one of the axes

internal energy total kinetic energy and potential energy of all the particles in a system

inversely proportional relationship a relationship of the form $y \propto 1/x$

inverted an inverted image is upside down compared with the object

ionisation process in which electrons split away from their atoms; some radiation is harmful to living cells because it is ionising

irradiation process in which an object is exposed to radiation; this does not make the object itself radioactive

isolated system physical system so far removed from other systems that it does not interact with them

isotopes atoms of an element containing the same number of protons but different numbers of neutrons

J

joule SI unit of energy, symbol J

K

kilogram SI unit of mass, symbol kg; 1 kg is the amount of mass equal to that of the standard platinum–iridium cylinder that is stored near Paris, France

kinetic energy energy an object has because of its movement; kinetic energy is greater for objects with greater mass or higher speed

L

latent heat energy needed for a substance to change state without a change in temperature (e.g., the latent heat of vaporisation is the energy needed to turn a sample of liquid water into gas)

lens piece of curved glass or plastic designed to refract light in a specific way

lever device that uses the moment of a force to move an object

light-dependent resistor (LDR) electric circuit component with a resistance that decreases as the intensity of the light falling upon it increases

light-emitting diode (LED) electric circuit component that glows when a potential difference is applied to it

limit of proportionality the point beyond which the extension of an elastic object is no longer proportional to the force applied

line of best fit line on a graph that most closely matches all the data points to show a trend or pattern

linear relationship relationship between quantities in which increases in one quantity result in proportional increases or decreases in the other quantity; a graph of a linear relationship produces a straight line

liquid state of matter in which the particles of a substance are close together and attract each other but have a limited amount of movement; a liquid has a definite volume but will spread out to fill its container

live wire conducting wire connection that carries the alternating current from the supply

longitudinal wave wave motion in which the vibrations of the particles of the medium are parallel to the direction of energy transfer (e.g., sound waves)

loudspeaker (moving-coil loudspeaker) electric circuit component that converts an electrical signal into sound waves

M

magnet object containing material that produces its own magnetic field and so will attract other magnetic materials such as iron

magnetic field area around a magnet or current-carrying wire, where there is a force on magnetic materials or current-carrying wires

magnetic flux density quantity that measures the amount of magnetic flux (field lines) in an area perpendicular to the direction of the magnetic flux; symbol B, unit tesla, T

magnification measure of how much larger an image is than the object – if the image is smaller than the object, the magnification is less than 1

magnitude size of a quantity

main sequence star star that fuses hydrogen in its core, in which the gravitational forces inwards are balanced by the pressure of the nuclear fusion reactions outwards; the Sun is a main sequence star

mass quantity that measures the amount of matter in an object; symbol m, unit kilogram, kg

mass number total number of neutrons and protons in the nucleus of one atom of an element

medium (*pl.* media) material through which light or other types of wave travel

metre SI unit of distance, symbol m; 1 m is defined as the distance travelled by light in a vacuum in $\frac{1}{299\,792\,458}$ s (you do not need to remember this definition)

micrometer device adjusted by a rotating screw thread, which measures small distances to high precision

microphone (moving-coil microphone) electric circuit component that uses the generator effect to convert sound waves into an electrical signal (variations in an electric current)

microwave electromagnetic radiation with a range of wavelengths longer than infrared but shorter than radio waves; used to cook food in microwave ovens, and for satellite communication

moment turning effect of a force – moment is increased by increasing the force or the perpendicular distance between the line of action of the force and the pivot; symbol M, unit newton-metre, Nm

momentum the product of mass and velocity of an object, symbol p, unit kilogram metres per second, kg m/s

motor effect interaction between a magnetic field due to a magnet and a current-carrying wire that causes a force on the wire and so causes movement of the wire

N

National Grid network of cables and transformers that links power stations to consumers across the country

nebula a huge cloud of dust and gas from which stars are formed

net decline the ratio of the final value of the activity of a radioactive substance to the initial value in a given number of half-lives

neutral wire conducting wire connection that allows electric charge to return to its source

neutron a particle inside the nucleus in the atoms of nearly all elements; neutrons have no electric charge

neutron star very small, very high-density star that is produced after a supernova explosion

newton SI unit of force, symbol N; 1 N is the force needed to give an object of mass 1 kg an acceleration of 1 m/s^2

Newton's first law if the resultant force acting on an object is zero, a stationary object will remain stationary and a moving object will keep moving at a steady speed in a straight line

Newton's second law a resultant force on an object produces an acceleration in the same direction as the force that is proportional to the magnitude of the force and inversely proportional to the mass of the object; in equation form $F = ma$

Newton's third law whenever two objects interact, the forces they exert on each other are equal, opposite and of the same type

Newton's laws of motion three fundamental physical laws that describe how objects and forces interact

newtonmeter device used to measure force

non-contact force force that acts at a distance between two objects that are separated (e.g., force due to an electric, gravitational or magnetic field)

non-linear relationship any relationship between two variables which when plotted on a graph does not produce a straight line

non-renewable resource source of energy used by humans that will eventually run out (e.g., fossil fuels are non-renewable fuels)

non-uniform motion movement in which the speed of an object changes

normal line at right angles to a boundary – used to draw ray diagrams

nuclear decay reaction in which the numbers of protons and/or neutrons in the nucleus of one or more atoms change

nuclear equation equation that uses symbols to show the elements involved in a nuclear decay, including the atomic numbers and mass numbers

nuclear model model of the the atom with a small central nucleus, surrounded by orbiting electrons

nucleus very small volume at the centre of an atom that contains all the protons and neutrons, and so concentrates nearly all the mass of an atom

O

ohm SI unit of electrical resistance, symbol Ω; a component with a resistance of 1 Ω allows a current of 1 A to flow when a potential difference of 1 V is applied

Ohm's law a resistor obeys Ohm's law if the current in the resistor (at constant temperature) is directly proportional to the potential difference across it

ohmic resistor any resistor that obeys Ohm's law at constant temperature

opaque material that does not allow light to pass through it, e.g., aluminium

orbit the path of an object about a star or planet, such as the path of a moon about a planet

order of magnitude description of a quantity in terms of powers of ten; e.g., a distance of 100 m (= 10^2 m) is two orders of magnitude larger than a distance of 1 m

oscilloscope device with screen to show how the amplitude and frequency of an input wave varies – also called a cathode ray oscilloscope

P

P-wave longitudinal wave produced by movements of the Earth's crust, which can travel through solids and the liquid centre of the Earth

parallel (circuit) electric circuit in which the current divides into two or more paths before combining again

particle model model in which all substances contain large numbers of very small particles (atoms, ions or molecules); it is used to explain the different properties of solids, liquids and gases

pascal SI unit of pressure, symbol Pa; 1 Pa of pressure arises when 1 N of force is applied over an area of 1 m^2

peer review process in which scientific experiments, writings and theories are checked and evaluated by other scientists

penetrating power measure of how far different types of radiation can pass into different types of material

period time taken for one complete cycle of an oscillation

permanent magnet object or material that produces its own magnetic field even if it is not within the magnetic field of another object

pivot point around which a lever or a seesaw turns

planet large ball of gas or rock travelling around a star; e.g., Earth and other planets orbit our Sun

plum pudding model early model of the structure of an atom, which suggested that an atom was a solid sphere of positive electric charge with negatively charged electrons in it

pole (magnetic) point in an object to or from which magnetic field lines point; magnetic poles always appear in pairs: one north, one south

potential difference (p.d.) a measure of the energy transferred per unit charge as charges move between two points in a circuit – also called the voltage between two points

potential energy energy associated with an object because of its position or the arrangement of the particles of the system; e.g., the amount by which a material is stretched (elastic potential energy)

power (of ten) number of times that ten is multiplied by itself in a quantity, e.g., ten to the power two = 10^2 = 10 × 10 = 100

power (energy transfer) the rate at which energy is transferred or the rate at which work is done; an energy transfer of 1 J/s is equal to a power of 1 W

precision how closely grouped a set of repeated measurements are

prediction statement that forecasts what would happen under particular conditions, based on scientific experiment and knowledge (e.g., if a ball is held above the ground and released, it will fall with an acceleration that is predictable based on knowledge of the Earth's gravity)

prefix letter added before the symbol for a unit to show how many powers of ten a quantity contains; e.g., 1 MW = 10^6 W = 1000 000 W, where the prefix M (mega) means 10^6

pressure pressure at any point is the force acting at that point divided by the area over which the force acts; pressure increases if force increases or area decreases

primary coil coil of a transformer across which the input potential difference is connected

principal axis line passing through the centre of a lens and perpendicular to the plane of symmetry of the lens; the principal focus of the lens is always on the principal axis

principal focus point at which parallel rays entering a lens either cross over or appear to have come from

proton a particle inside the nucleus in all atoms; protons have a positive electric charge

protostar cloud of gas pulled together into a spherical volume by its own gravity, in which the compressed gas gets hotter and starts to glow

R

radiation energy given out in the form of electromagnetic waves or as moving particles; e.g. in radioactive beta decay a nucleus emits high-speed electrons, and the Sun radiates electromagnetic waves including visible light

radioactivity process in which particles or energy are produced by the reactions of unstable atomic nuclei

radiotherapy the use of ionising radiation to kill cancer cells or reduce the size of a tumour

radio waves electromagnetic radiation with a range of wavelengths longer than microwaves; used for long-distance communication

random (radioactive decay) process in which the time of each particular event cannot be predicted, although a trend or average can be measured across many events; e.g., the decay of a radioactive element

random error estimated amount by which a measurement or calculated quantity is different from the true value, due to results varying in unpredictable ways

rarefaction a region of a sound wave where the particles are further apart

ray diagram line diagram showing how rays of light travel

reaction time time it takes a vehicle driver to respond to a danger on the road; the 'thinking distance' is the distance the vehicle travels during this reaction time

real image image that can be projected onto a screen or a light sensor

red giant star in which all the hydrogen has fused to form helium, at which time the star expands to a much greater size and helium starts fusing to form other elements

red super giant very large and bright star that has a diameter more than 100 times larger than the Sun

red-shift an increase (shift) in the wavelength of spectral lines of light from most distant galaxies, moving the lines to the red end of the spectrum

reflection process in which a surface does not absorb any energy, but instead bounces it back towards the source; e.g., light is reflected by polished surfaces

refraction change of direction of a wave when it hits a boundary between two different media at an angle; e.g., when a light ray passes from air into a glass block

renewable resource source of energy that can be replaced or reused over a short time, e.g., biofuels from crops that can be grown again from seed

repulsion force that acts between two objects that tends to push them further apart, e.g., the force that arises between two positive electric charges

resistance ratio of the potential difference across an electrical component to the current through the component; symbol R, unit ohms, Ω

resistor electric component that produces a desired amount of resistance to the current within it when a potential difference is applied across it

resolution (forces) splitting a single force into two components acting in different directions, to simplify a calculation

resultant force the single force that would have the same effect on an object as all the forces that are acting on the object

S

S-wave transverse wave produced by movements of the Earth's crust, which can travel through solids but not through the liquid centre of the Earth

satellite any natural or artificial object orbiting around a larger object

scalar quantity measurable quantity that has only a magnitude, not a direction (e.g., mass)

scattering process in which particles or electromagnetic waves are deflected or reflected in a number of different directions; e.g., the reflection of light off a rough, matt surface (diffuse reflection)

second SI unit of time, symbol s; defined as the duration of 9 192 631 770 periods of the radiation corresponding to the transition between the two hyperfine levels of the ground state of the caesium-133 atom (you do not need to remember this definition)

secondary coil coil of a transformer across which an output potential difference is induced

seismic wave wave of energy produced by an earthquake or other movements of the Earth's crust, which can travel through the Earth or across its surface

series (circuit) electric circuit in which all components are connected one after the other in a single line

SI unit standard units of measurement, one per quantity, used by all physicists; all SI units are derived from seven 'base' units that have precise definitions

sievert SI unit of radiation dose, symbol Sv

significant figures digits within a measured quantity that have meaning; e.g., a measurement made using a 30 cm long ruler with divisions marked in millimetres can only have three significant figures, such as 17.4 cm; it is meaningless to state 17.42 cm because the ruler is not that precise (note: it may be possible on some rulers to estimate a measurement to the nearest 0.5 mm)

slope see gradient

soft iron core laminated core of soft iron around which the coils of a transformer are wound; the current in the primary coil causes a magnetic field in the soft iron core

solar system a star and all the planets, moons, comets and other objects that orbit the star; our Solar System includes the Sun and eight major planets

solenoid a coil of current-carrying wire that generates a magnetic field

solid state of matter in which the particles are held together in a fixed structure by bonds

specific heat capacity the energy needed to raise the temperature of 1 kg of a substance by 1 °C; symbol c, unit J/kg °C

specific latent heat the energy needed to change 1 kg of a substance completely from one state to another state without any change in temperature; symbol L, unit J/kg

specific latent heat of fusion the energy needed to change 1 kg of a substance completely from solid to liquid without any change in temperature

specific latent heat of vaporisation the energy needed to change 1 kg of a substance completely from liquid to gas without any change in temperature

specular reflection reflection from a smooth surface in a single direction

speed the distance travelled by an object per unit of time; unit metres per second, m/s

speed of light speed at which electromagnetic radiation travels through a vacuum; 300 000 000 m/s

spring constant quantity that tells you how much an object (such as a spring) will stretch by if a force is applied to it, as long as the object obeys Hooke's law; symbol k, unit newtons per metre, N/m

star ball of gas in space with enough mass that the gravity pulling the gas towards its centre is enough to cause nuclear fusion reactions within it

state of matter form that particles of a substance take depending on temperature; different states include solid, liquid and gas

static electricity electric charge that has accumulated on or within an object, causing the object to produce an electric field

step-down transformer transformer that changes an alternating potential difference across the primary coil to a lower potential difference across the secondary coil

step-up transformer transformer that changes an alternating potential difference across the primary coil to a higher potential difference across the secondary coil

stopping distance total distance a vehicle travels before coming to a complete stop; stopping distance = thinking distance + braking distance

stretching process in which a force pulls the particles of a material further apart, causing the material to extend

supernova exploding star, caused when nuclear fusion stops within a star with a large enough mass; the mass of the star collapses inwards rapidly, and the colliding mass causes an explosion

systematic error consistent amount by which a measurement differs from the true value each time it is measured, due to the experimental technique or the set-up; e.g., an instrument not correctly calibrated, or background radiation in the measurement of radioactive decay

T

temperature measure of the hotness or coldness of an object or environment

tension force that pulls or stretches

terminal velocity constant velocity that occurs when the gravitational force acting downwards on a body falling through a fluid is exactly balanced by the upwards force due to the resistance of the fluid

tesla SI unit of magnetic flux density, symbol T

thermal conductivity measure of the ability of a material to conduct thermal energy from a hotter place to a colder place

thermal energy internal energy present in a system due to its temperature, which itself is due to the random motion of the particles within the system

thermistor electric circuit component with a resistance that decreases as its temperature increases

thermometer device used to measure temperature

thinking distance distance a vehicle travels during the time it takes a vehicle driver to respond to a danger on the road

time period time taken for one complete oscillation of a wave, symbol T

tracer a radioactive substance that is put into the body or fluid (such as in a pipe), so that the path of the substance can be followed by monitoring the radiation it emits

transformer device used to increase (step up) the potential difference of an alternating signal or decrease it (step down)

translucent material that allows light to pass through it but diffuses the light (scatters parallel light rays in different directions) so that clear images cannot be seen through it

transmission movement of energy or information from one position to another; e.g., microwaves are used to transmit mobile phone signals to and from an aerial (signal mast)

transparent material that allows all light to pass through it without scattering; e.g., clear water is transparent, and images can be seen through it

transverse wave wave motion in which the vibrations of the particles of the medium are perpendicular to the direction of energy transfer (e.g., water waves or electromagnetic waves)

U

ultrasound sound waves that have a frequency too high for humans to hear

ultraviolet radiation electromagnetic radiation with a range of wavelengths shorter than visible light but longer than X-rays; emitted in particular by the Sun

uncertainty the uncertainty of a measurement is half the range of values recorded, from maximum to minimum; uncertainty is reduced when accuracy and precision are increased

uniform motion movement of an object in a straight line at a constant speed

V

vector quantity measurable quantity that has both a magnitude and a direction (e.g., velocity)

velocity speed at which an object is moving in a particular direction; symbol v, unit metres per second, m/s

velocity–time graph graph with velocity on the y-axis and time on the x-axis; the gradient of a velocity–time graph is equal to the acceleration; the area under a velocity–time graph is equal to the displacement

Vernier calipers device adjusted by moving one calibrated scale over another, which measures small distances to high precision

virtual image image that can be seen but cannot be projected onto a screen (mirrors and concave lenses form virtual images)

visible light electromagnetic radiation with a range of wavelengths shorter than infrared but longer than ultraviolet; detectable with the human eye

volt SI unit of potential difference, symbol V; 1 V is the potential difference between two points on a conducting wire when an electric current of 1 A dissipates 1 W of power between those points

voltage see potential difference

voltmeter device that measures potential difference

W

watt SI unit for power, symbol W; a power of 1 W is equal to 1 J of energy transferred in 1 s

wave a disturbance (oscillation) that transfers energy or information from one point to another

wavelength distance between a point on one wave to the equivalent point on the adjacent wave

weight measure of the force of gravity on an object

white dwarf star nearing the end of its life, where all of the nuclear fuel has been used up; a white dwarf slowly cools until it becomes a black dwarf

work done work is done when a force acts on an object and the object moves along the line of action of the force; symbol W, unit joules, J or newton-metres, Nm

X

X-ray ionising electromagnetic radiation with a range of wavelengths shorter than ultraviolet and can have similar wavelengths to gamma rays; used in X-ray photography to generate pictures of bones or teeth and in CT scans

Index

ABS (anti-lock braking system) 154

absorption

   of wave energy 202, 226–7, 232, 234–5

   lines 111

acceleration 140–1, 145–51, 158–61

   definition 146

   due to gravity 146, 291

   investigating 160–3

accelerometer 160

activity of radioisotope definition 112

air bags 141

air resistance 140–2, 146, 154, 158

algebraic equation (maths skill) 76–7

alpha decay 109, 112–3, 116

alpha particle 109, 112–6, 132–3

   scattering experiment 132–3

alternating current (a.c.) 46, 66

alternative energy (see renewable energy)

americium-241 116

ammeter 52–6

amplitude 191–2

analysing 31, 161, 178

angle

   of incidence 203, 205

   of reflection 203, 205

   of refraction 203, 205

anomalous 40–1

anomaly 178

Archimedes 86–7

argon 59

artificial satellite 275

atmospheric pressure 174–5

atom 90, 100, 108–9, 132–3

atomic

   bomb 123

   number 109, 111, 116–17

   radius 110

   structure 108–11, 132–3

   symbol 108–9

Aurora Borealis 281

balanced forces 154–6

bar chart 40–1

battery 52

becquerel (Bq) 112

Becquerel, Henri 11

beta

   particle 109, 115

   decay 109, 113, 116–17

Big Bang 275, 289

black body 232–3

black dwarf 284

black hole 285, 290

blue whale 207

Bluetooth 224

Bohr, Niels 133

boiling 88–9, 94–5

bonds 84, 89, 91, 100–1

boron 129

brachytherapy 125

braking distance 166

brushes (generator, alternator and motor) 254–5, 262–3

buzzer 56

camera 190

cancer 122–7, 216–7

carbon 287

carbon dioxide 235

carbon-15 113, 117

CAT (computerised axial tomography) scan 124, 217

cathode ray oscilloscope (CRO) 195

cell (electrical) 46, 52

centre of mass 157, 169

Chadwick, James 133

chain reaction 128

changes of state 82–3, 88–9, 91, 94–5, 100–1, 102–3

charge (electrical) 46–52, 71, 74–5

Charon 277

chemical equation 108, 116

chemical reaction 108

circuit symbols 52

circuit-breaker 46–7, 66

circular motion 141, 147

circular orbit 278–9

cloud cover 235

coal 13, 34, 37

cobalt 245

cobalt-60 125

collision 163

inelastic 165

colour 226–7

commutator 243, 254, 262

compact fluorescent light 218

compass 246, 260

plotting 245

compression 176–7, 194, 206, 210, 212, 257

of gas 99

condensing 88–9

conduction 12, 26–7, 30–1, 219

conductor (electrical) 48

conservation

of charge 116–7

of energy 28–9, 99, 170

of mass 88, 116–17

of matter 108–9

contact force 140, 142

continuous data 40–1

control circuits 64–5

control rod, boron 129

convection 12, 26–7, 30–1

cooling curve

   stearic acid 102

   wax 103

core (of Earth) 211, 246

correlation 40–1

cosmic rays 114, 122

coulomb, definition 52

Crab Nebula 285

crumple zone 164

CT (computerised tomography) scan – see
   CAT (computerised axial tomography)
   scan current

   electrical 46–7, 52–69, 72–5

   induced 258, 260–1

current 46–7, 52–69, 72–5, 247–9

curve of best fit (maths skill) 118

cylindrical magnet 256

d.c. generator 243, 258–9, 261

dark energy 275, 291

dark matter 275, 291

daughter nucleus 128

deceleration 146, 183

deforestation 235

deformation

   elastic 176–7

   plastic 176–7

Democritus 132

density 82, 84–7, 173, 207, 237

   of water (anomalous expansion) 91

dependent variable 40–1

deuterium 109

diode 52–3, 65

direct proportion 76–7

discrete data 40–1

displacement 143, 149, 192

dissipation of energy 26–7

distance–time graph 140, 144–5

DNA 122, 286

double insulation 67

drag 140, 146, 154

drawing graphs 102

drop tower 181

dry ice 88

dwarf planet 275, 277

dynamics trolley 160

dynamo 261–62

eardrum 206–7

Earth 274–9, 290, 292

   structure 211

earth (wire) 66

earthquake 196, 210–11

echo sounding 191, 199

efficiency 28–9, 34, 265

effort force 170–1

elastic potential energy 14–15, 177

electric

  bell 250

  current 46–7, 52–69, 72–5, 247–9

  current (a.c.) 263

  field 50–1

  motor 243, 254–5, 261

  spark 51

  conductor 48

electrical

  energy 32–3

electrical power 47, 70–3, 75, 77, 265, 266

  appliance 67

  battery 46

  cell 46, 52

  charge 46–52, 71, 74–5

electricity 46–77

  transmission 68–9

electromagnet 242–3, 248, 250–1

electromagnetic

  spectrum 212–13, 216–19

  wave 111, 212–13, 190–1, 197, 213, 216–19

electromagnetism 242–67

electron 108–11, 132–3

  as charge carrier 53

  energy level 111

  transfer 49

electrostatics, law of 49

element 275, 283–7

  formation 286–7

emergent ray 203

emission (of wave energy) 220, 232–3, 234–5

energy resources

  bio-fuel 32–3, 41

  coal 13, 34, 37

  economic considerations 35

  ethical considerations 35

  fossil fuels 12–13, 34–5, 235

  gas 13, 34–5

  geothermal power 32–3, 41

  global energy supplies 34–5

  hydroelectric power 32–3, 41

  non-renewable 32–3

  nuclear power 32–3

  oil 13, 34

  renewable 12–3, 32–5, 41

  social considerations 35

  solar power 32–3, 41

  tidal power 32–3

  wave power 32–3

  wind power 32–3, 41

energy transfer 12–37, 195–7, 261

  conduction 12, 26–7, 30–1, 219

  convection 12, 26–7, 30–1

  radiation 12, 26–7, 30–1, 220–1

energy

    conservation 28–9, 99, 170

    dissipation 26–7

    elastic potential 14–15

    electrical 32–3

    energy, efficiency 28–9, 34

    gravitational potential 14–17, 38–9

    internal 83, 90–1, 97, 100–1

    kinetic 16–19, 38–9

    store 24–5

    stored 14–17

    thermal (heat) 12–13, 22–3, 91, 100–1

    transfer 70–1, 77, 88–9, 100–1

    wave 196–7, 201, 202, 208, 212–14

    work done 18–21, 99, 170

environmental tracer 121

equations, rearranging (maths skill) 76–7

equilibrium 282

equivalent

    circuit 57

    resistance 57

estimating (maths skill) 182–3

Eureka can 86

evaluating 25, 31

evaporating 88–9

expanding Universe 289

extension 176–7

eye 190

Faraday, Michael 247, 258, 260

Fermi, Enrico 182

fetus 208

Feynman, Richard 37

field, gravitational 279

filament lamp 58–9

filter (light) 226–7

Fleming's left-hand rule 249, 254

floating 173

fluid pressure 172–3

fluorescence 218

focal length 228–31

force

    and work done 18–19

    as a push or pull 142

    components 157

    contact 140, 142

    effort 170–1

    load 170–1

    multiplier 170

    non-contact 140, 142–3, 163, 290

    normal contact 49, 142, 155, 163

    on a conductor 266–7

    pairs 162

    resultant 155–8

    thrust 142, 156

    turning 168–9

forces

    balanced 154–6

    unbalanced 146, 156, 180–1

formulae, using (maths skill) 76–7

fossil fuels 12–13, 34–5, 235

free-body diagram 156, 162

freezing 88–9, 102

frequency 190–3, 200–1, 206–9, 213–15, 218, 222, 224, 236–7, 275

  of mains electricity 66

  table 40–1

friction 26–7, 156, 148, 167

fuel rod 129

fulcrum 170

fuse 46, 52, 66–7

fusion, cold 131

galaxy 274–5, 280, 289–93

Galileo 146

gamma

  camera 126

  decay 113, 117

  ray 109, 115, 124, 126, 197, 213, 216–17

gas 13, 34–5, 82–5, 96–101

  pressure 82–5, 96–9, 257

  volume 98

gears 171

Geiger, Hans 132

generator, d.c. 243, 258–9, 261

generator effect 258, 261–3

geothermal power 32–3, 41

global energy supplies 34–5

global warming 34

gold foil 132–3

gradient of graph 145, 179

  tangent 145

granite 114

graph skills 77, 179

graphite moderator 129

gravitational

  field 279

  field strength 14–5, 152–3, 173, 274, 291

  potential energy 14–7, 38–9

gravity 140, 274–5, 278–82, 290–1

Green, Andy 150

greenhouse gases 235

half-life 118–19

  definition 118

halogen hob 219

haybox 31

headphones 257

heat

  energy (see thermal energy)

  insulator 30–1

  radiation 12, 26–7, 30–1, 220–1

Helios 2 158

helium 109, 111–12, 130–1, 280–1, 283–4, 286–7

hertz (Hz) 190

histogram 40–1

Hooke's law 178–9

Hooke, Robert 178

Hubble, Edwin 289

hydroelectric power 32–3, 41

hydrogen 109, 130–1, 280–1, 283–4, 286–7

  bomb 131

hypothesis forming 30

image 204

  real 228, 230–1

  virtual 229, 231

incident ray 202, 205

independent variable 40–1

induced current 258, 260–1

induced potential difference 258–60

inertia 159, 180

infrared

  absorption 220

  detector 221, 233

  emission 221

  radiation 191, 196–7, 213–14, 219–21, 230, 232–5

insulator

  electrical 46–7, 48, 51

  heat 30–1

internal energy 83, 90–1, 97, 100–1

International Space Station 279

interpreting graphs 103

investigation skill

  analysing 131, 161, 178

  drawing graphs 102

  evaluating 25, 31

  graph skills 179

  hypothesis forming 30

  interpreting graphs 103

  planning 24, 161

  using data 24

iodine-123 125–6

ionisation 111

ionising radiation 111, 217

ionosphere 225

iron 242–3, 287

irradiation 122–3

isotope 109–10, 117

  definition 110

$I–V$ graph 59, 64–5

Jupiter 14, 276–7

kicking wire 249

kilogram 152–3

kinetic energy 16–9, 38–9, 167

  of particles 83, 88–92, 96–7

lamp 52

latent heat 103

law of magnets 212, 214

LDR (light dependent resistor) 52, 53, 64–5

lead-208 112

LED (light emitting diode) 52

lens

  concave 191, 228–9

  convex 191, 228–31

lever 170

light dependent resistor 52, 53, 64–5

light gates 160

light year 274

lightning 48, 51, 75

lightning, conductor 51

limit of proportionality 176–7

lines of best fit (maths skill) 40–1

liquid 82–5, 100–1

live (wire) 66–8

load force 170–1

longitudinal wave 190, 194–5, 210, 212–13, 256–7

loudspeaker, moving-coil 256–7

lubrication 26

machine 170

maglev 244

magnesium 287

magnet

    bar 244

    cylindrical 256

    electromagnet 242–3, 248, 250–1

    permanent 242, 245

magnetic

    attraction 244–5

    effect of current 242–3, 247

    field 244–7, 260–1

    field around a solenoid 248, 252

    field around a wire 247

    field lines 244–5

field of Earth 246

field, radial 255

field strength 252

flux density 252

induction (induced magnetism) 245, 250

    pole 244

    repulsion 244

    switch 251

magnification 231

Mars 276–7, 292

Marsden, Earnest 132

mass 152–3

    inertial 158–9

    number 109, 111, 116–17

mean (maths skill) 40–1, 178

median 40–1

medical

    tracer 121, 126, 216–17

    imaging 126, 216–17

melting 88–9, 94–5, 103

mercury 58

Mercury (planet) 276–7

methane 235

micrometer 87

microorganisms 122

microphone, moving-coil 195, 262

microwave oven 222

microwave radiation 196–7, 213–14, 222–3

Milky Way 183, 274, 280, 290, 292–3

mobile phone 197, 22

   charger 264

mode 40–1

moderator 129

molecule 82, 88–91

moment 168–9

momentum 141, 164–5

   conservation of 165

   definition 164

Moon 278–9, 290, 292

moons 275, 277

Morse code 197

motion

   calculations 150–1

   equations of 150–1

   in a circle 141, 147

   uniform accelerated 150–1

   vertical 151

motor effect 249

moving-coil microphone 262

mutation, gene or cell 122

National Grid 68–9, 243

natural satellite 278

nebula 275, 280, 284–5, 287–8

neon 287

Neptune 276–7

neutral (wire) 66–7

neutron 108–11, 130, 133

radiation 112, 128

   star 285

newton (N) 142, 152–3

Newton, Isaac 178

Newton's first law 154–5, 180

Newton's second law 158–61, 181, 291

Newton's third law 162–3

newtonmeter 152

nickel 245, 287

nitrogen 59, 287

nitrogen-15 113, 117

nitrogen oxides 235

non-contact force 140, 142–3, 163, 290

non-ohmic resistor 59

non-renewable energy resources 32–3

normal (line) 202–3, 205

normal contact force 49, 142, 155, 163

Northern Lights (see Aurora Borealis)

nuclear

   equation 116–17

   fission 128–9

   fusion 130–1, 275, 280–7

   power 32–3

   radiation 126

   power station 129

   radiation, definition 112

   reaction 109, 128

nucleon 110–11, 113, 116

nucleus 110, 132–3

Oersted, Hans Christian 247, 260

Ohm's law 46–7, 53, 58, 63, 76–7

oil 13, 34

opaque 226

optical fibre 203

orbit 276, 278–9

order of magnitude 183

osmium 85

outlier 41

oxygen 287

P wave 210–11

parachute 140–1, 154

parallel circuit 54–5, 62–3

particle

    model 82–5, 89–90, 96, 98, 175

    subatomic 109

pascal (Pa) 82, 172–3

peer review 122–3

perfect reflector 202

periodic table 109, 117

permanent magnet 242, 245

perpetual motion machine 38–9

phytoplankton 202

pivot 168–70

planets 275–8

planning 24, 161

plug, three-pin 66

plum pudding model 132

Pluto 276–7

plutonium 128

potassium-40 122

potential difference 52–69, 71–77, 264–5

    induced 258–60

potential energy, of particles 90

power 13, 20–1

    electrical 47, 70–3, 75, 77

pressure

    atmospheric 174–5

    definition 172

    fluid 172–3

    gas 82–5, 96–9, 257

primary coil 264

principal axis 228–31

principal focus 228–31

principle of moments 169

prism 226

proportional reasoning (maths skill) 134–5

proton 108–11, 133

protostar 280

pumped storage hydropower 15

radar 223

radiation

    background 114, 121

    balance 234–5

    dose 114–15, 126–7, 216–7

    hazard symbol 120

    infrared 191, 196–7, 213–14, 219–21, 230, 232–5

microwave 196–7, 213–14, 222–3

nuclear 122–3, 127

nuclear, absorption 115

nuclear, properties 115

ultraviolet radiation 191, 197, 213, 218–9, 282

radiator (domestic) 93

radio telescope 232

radio wave 196, 213, 215, 224–5

radioactive

contamination 120

decay 109, 112–13, 116–19, 217

half-life 118–19

radioisotope 109, 118–21, 124, 131

radiotherapy 124–7

radium 110

random motion, particle 96

range of hearing 199, 206

rarefaction 194, 206, 210, 212, 256

ratio (maths skill) 134–5

ray diagram 191, 202–5, 215, 228–9

reaction time 141, 166

rearranging equations (maths skill) 38–9, 76–7, 237, 266–7

receiver 224–5

red giant 284

red super-giant 285, 287

red-shift 275, 288–9

reflected ray 202, 205

reflection 190–1, 202–5, 211, 214, 224–7, 234–15

diffuse 204

specular 204

refracted ray 202, 205

refraction 191, 202–3, 205, 208–9, 211, 214–15, 224–5

relationship

between variables 76–7

linear 76–7

relay 251

renewable energy 12–13, 32–5, 41

resistance 53, 56–65, 72, 75–7

measuring 56, 58

resistor 46–7, 52

non-ohmic 59

resistors in parallel 55, 63

resistors in series 54, 57, 63

resultant force 155–8

retroreflector 204

right-hand grip rule 247

ripple tank 200

road safety 166–7

rocket 96

rounding numbers (maths skill) 182–3

Rutherford, Earnest 132–3

S wave 210–11

satellite communication 224–5

satellite phone 225

Saturn 276–7

scalar 143

scale (maths skill) 292

scatter diagram 40–1

season 274

seat belts 141

secondary coil 264

seismic wave 191, 210–11

seismometer 210

semi-conductor 64–5

series circuit 54, 62–3

shadow zone 211

sievert (Sv) 115, 127, 216

significant figure (maths skill) 39, 87

silicon 287

sinking 173

sketch graph 148

skin cancer 218

slip rings (motor and alternator) 255, 263

smoke alarm 116, 119

solar

   flare 281

   power 32–3, 41

   wind 281

Solar System 275–8, 292

solenoid 248, 252, 258–9

solid 82–5, 100–1

sound

   wave 190–1, 206–7, 213, 256–7

   speed of 198–9, 207

space 274–93

specific heat capacity 22–5, 92–3

specific latent heat

   of fusion 83, 95–6

   of vaporisation 83, 95–6

speed 140–1, 143–5

   average 140, 144

split-ring commutator 243, 254, 262

spring 176–7

   constant 177

standard form (maths skill) 292–3

standing wave 201

star 130, 274–5, 277, 290

   life cycle 275, 282–5

   main sequence 282–3

states of matter 84, 100–1

static electricity 46–9

stearic acid, cooling curve 102

steel 245

step-down transformer 243, 264–5

step-up transformer 243, 264–5

Stonehenge 170

stopping distance 141, 166–7

stored energy 14–17

stores of energy (see energy stores)

sublimating 88–9

sulfur 287

Sun 275–8, 280–3, 290

sun screen 218

sun tan 218

superconductor 53

supernova 285, 287, 291

switch 52

technetium-99m 119, 216

temperature 12–3, 22–3, 82–3, 92–103

   of Earth 234–5

terminal velocity 141

thermal

   energy (heat) 12–13, 22–3, 91, 100–1

   image 27, 219, 232

   insulation 26–7

thermistor 47, 52, 64–5

thermogram 26–7

thinking distance 166

Thompson, J. J. 132

three-pin plug 66

thrust force 142, 156

Thrust SSC 150

ticker-timer 160

tidal power 32–3

time period (wave) 192

torque 170

transfer of energy 70–1, 77, 88–9, 100–1

transformer 68–9, 243, 264–5, 267

   step-down 243, 264–5

   step-up 243, 264–5

translucent 226

transmission (of wave energy) 202, 226–7

transmitter 224–5

transparent 226

transverse wave 190, 194, 211–13

tsunami 196

tumour 125, 127, 217

tungsten 59

tungsten filament lamp 58–9

turning force 168–9

TV 224–5

   remote control 219

ultrasound 190, 199, 208–9

ultraviolet radiation 191, 197, 213, 218–19, 282

unbalanced forces 146, 156, 180–1

uncertainty, of measurement 87

Universe 275, 280, 286–93

upthrust 173

uranium-235 128–9

uranium-238 110–11, 113, 116, 129

Uranus 276–7

using data 24

using formulae (maths skill) 76–7

vacuum 190, 195, 213, 217

Van de Graaff generator 49, 75

variable resistor 52

vector 143, 157, 162–3

   diagram 157

velocity 141, 143, 146–9, 182–3

   terminal 141

velocity–time graph 148–9

Venus 276–7

Vernier callipers 87

vibration 195–6, 201, 206–7, 222, 256, 262

visible light 190–1, 226–7, 234

visible spectrum 111, 212–13, 226–7, 288–9

vitamin D 218

voltmeter 52–3, 56

volume, of a regular object 183

water wave 194, 196, 198, 200–1, 212–13

Watt (definition as the unit of power) 20

wave power 32–3

waves 190–237

   as information carrier 196–7, 224–5

   electromagnetic 190–1, 197

   energy 196–7, 201, 202, 208, 212–14

   equation 193, 209, 237

   frequency 190–3, 200–1, 206–9, 213–15, 218, 222, 224, 236–7, 275

   front 214–15

longitudinal 190, 194–5, 210, 212–13, 256–7

P 210–1

radio 196, 213, 215, 224–5

S 210–11

seismic 191, 210–11

sound 190–1, 206–7, 213, 256–7

speed 191, 198–201, 203, 237

standing 201

time period 192

transverse 190, 194, 211–13

   water 194, 196, 198, 200–1, 212–13

wavelength 191–3, 200–1, 213, 237, 288–9

weight 152–3

weightlessness 153

wheelbarrow 170

white dwarf 284

wind power 32–3, 41

work done 18–21, 99, 170

X–rays 122, 124–5, 196–7, 208, 213, 216–17, 282